Undergraduate Lecture Notes in Physics

Series Editors

Neil Ashby, University of Colorado, Boulder, USA

William Brantley, Department of Physics, Furman University, Greenville, USA

Matthew Deady, Physics Program, Bard College, Annandale-on-Hudson, USA

Michael Fowler, Department of Physics, University of Virginia, Charlottesville, USA

Morten Hjorth-Jensen, Department of Physics, University of Oslo, Oslo, Norway

Michael Inglis, Department of Physical Sciences, SUNY Suffolk County Community College, Selden, USA

Barry Luokkala ⓘ, Department of Physics, Carnegie Mellon University, Pittsburgh, USA

Undergraduate Lecture Notes in Physics (ULNP) publishes authoritative texts covering topics throughout pure and applied physics. Each title in the series is suitable as a basis for undergraduate instruction, typically containing practice problems, worked examples, chapter summaries, and suggestions for further reading.

ULNP titles must provide at least one of the following:

- An exceptionally clear and concise treatment of a standard undergraduate subject.
- A solid undergraduate-level introduction to a graduate, advanced, or non-standard subject.
- A novel perspective or an unusual approach to teaching a subject.

ULNP especially encourages new, original, and idiosyncratic approaches to physics teaching at the undergraduate level.

The purpose of ULNP is to provide intriguing, absorbing books that will continue to be the reader's preferred reference throughout their academic career.

Joel L. Schiff

Basic Mathematical Insights into Astrophysics

 Springer

Joel L. Schiff
Department of Mathematics
University of Auckland
Auckland, New Zealand

ISSN 2192-4791   ISSN 2192-4805 (electronic)
Undergraduate Lecture Notes in Physics
ISBN 978-3-031-79099-7   ISBN 978-3-031-79100-0 (eBook)
https://doi.org/10.1007/978-3-031-79100-0

© The Editor(s) (if applicable) and The Author(s), under exclusive license to Springer Nature Switzerland AG 2025

This work is subject to copyright. All rights are solely and exclusively licensed by the Publisher, whether the whole or part of the material is concerned, specifically the rights of translation, reprinting, reuse of illustrations, recitation, broadcasting, reproduction on microfilms or in any other physical way, and transmission or information storage and retrieval, electronic adaptation, computer software, or by similar or dissimilar methodology now known or hereafter developed.
The use of general descriptive names, registered names, trademarks, service marks, etc. in this publication does not imply, even in the absence of a specific statement, that such names are exempt from the relevant protective laws and regulations and therefore free for general use.
The publisher, the authors and the editors are safe to assume that the advice and information in this book are believed to be true and accurate at the date of publication. Neither the publisher nor the authors or the editors give a warranty, expressed or implied, with respect to the material contained herein or for any errors or omissions that may have been made. The publisher remains neutral with regard to jurisdictional claims in published maps and institutional affiliations.

This Springer imprint is published by the registered company Springer Nature Switzerland AG
The registered company address is: Gewerbestrasse 11, 6330 Cham, Switzerland

If disposing of this product, please recycle the paper.

We are a way for the Universe to know itself… Carl Sagan

'The Mind of the Universe' created by the author with Midjourney.

Preface

The Universe is constantly sending us information about itself in the form of electromagnetic radiation. Some of it is visible, and some is not. But just how do we decipher this radiation and turn it into useful information? That is the subject of this book.

The best language we have to decode what the Universe is telling us is mathematics. Yes, we can say that the planets revolve around the Sun but if you want to send a spacecraft to orbit Mars, then that is not enough. And often, the mathematics is very elegant such as Newton's Law of Universal Gravitation, Kepler's Laws, or even Einstein's Field Equations. Through mathematics we can make predictions about the fate of the Universe and even something about its origins. What we cannot explain with mathematics is why there is a Universe at all, which is a question to be debated by philosophers. But given the fact that we have one, let's study as much as we can about it using the language of mathematics.

It is also rather remarkable that we have a language to describe the workings of the Universe. Somehow, from a language that started out with counting on our fingers, and went on to describe ideal perfect circles and triangles, whose interior angles sum to exactly 180 degrees, we have ended up describing the orbits of planets around the Sun, the mass of a black hole at the center of the Milky Way, or the ultimate fate of the Universe. And all of this is often achieved involving such quantities as π and e which have never-ending decimal expansions and come out of purely theoretical considerations. What does the ratio of the circumference of a perfect circle to its diameter have to do with the density of matter in the Universe? Strangely enough, it does.

And lest the reader think that all the important calculations concerning the workings of the Universe are only historical in nature, from the likes of Newton, Kepler, and Einstein, then that would be an incorrect assumption. Yes, it has been known since the observations of Edwin Hubble in 1929 that the Universe is expanding, but it is only in modern times (1998) that it has been discovered that the expansion of the Universe is actually accelerating, and has been for some billions of years. A consequence of this acceleration is that the Universe will continue to expand over eons of time and ultimately it is literally lights out, forever more. Of course, there could be future developments that might indicate a different scenario for the ultimate fate of the Universe as the cause of the acceleration is not at all well understood. That is why the study of this issue is so fascinating because the fate of the Universe hangs in the balance.

And the good news is that to get started on this journey of our exploration of the Universe, one can get deeply into the subject with just undergraduate-level mathematics. We will even be able to follow in the footsteps of Albert Einstein and perform the same landmark calculation he did in 1915 in order to calculate the precession of the orbit of Mercury. And of course, we get the same value, the famous 43 arcseconds per century. We will also show that the notion of time, which most people think of as immutable, is actually flexible and we will demonstrate this fact using nothing more than the Pythagorean theorem.

This latter fact is a consequence of Einstein's Theory of Relativity, and while it has a formidable reputation, there are many aspects of the theory that are readily accessible to an undergraduate. And one of the most remarkable aspects that we will delve into is the fact that without taking into account the fact of time dilation, there would be no such thing as GPS (Global Positioning System) and the world of navigation would be very much different. Indeed, one of the many consequences of the Theory of Relativity that we will examine is the extra constant factor that Einstein put into his equations so that the Universe would remain static. However, when it was found that the Universe was actually expanding, he considered this his 'greatest mistake'. But we should all make such mistakes because this extra constant term in recent years has made a comeback and could account for what is driving the acceleration of the Universe and to the final end of everything. In this one little constant, we are staring at the fate of the Universe and its likely ultimate end.

Yet so much is still unknown about the Universe that it is crying out for further investigation. In spite of the Universe not being very amenable to study, like a cockroach or electron in a laboratory, astronomers over the centuries have turned the Heavens into a laboratory, and the reader will be quite surprised by how much is already known about how it all works. Indeed, prepare to be amazed by just how much can be learned about our celestial habitat we call the Universe.

Auckland, New Zealand Joel L. Schiff
2024

Acknowledgements

The author wishes to acknowledge fruitful discussions with Jenni Adams, University of Canterbury; Tamara Davis, University of Queensland; and most especially, my good friend Jonathan Park, who read the entire manuscript, verified (or sometimes not) countless calculations, and made many valuable contributions to the text as well as contributing many superb images. I really cannot thank him enough. I also wish to thank Katy Metcalf who also provided many of the fine graphics. Any remaining errors are owned by the author.

Contents

1	**Measurement**	1
	Scientific Notation	1
	SI Units	2
	Constants	3
	What's the Angle?	4
	Solid Angles	6
	Light-Years/Parsecs	9
	Mass	10
	Density	10
	Temperature	11
	Logarithms	12
	Brightness/Stellar Magnitudes	12
	Galactic Orbital Velocities	14
	Method of Least Squares	15
	Standard Deviation	17
	Power Laws	19
	Mass-Luminosity Relation	20
	Main-Sequence Lifetimes	21
2	**Down to Earth**	23
	Size of Earth	23
	Age of the Earth	23
	Orbiting Earth	28
	Escaping Earth	29
	Mass of Earth	29
3	**Let There Be Light**	31
	Electromagnetic Radiation	31
	Frequency	33
	Energy	34

xi

Blackbody Radiation 35
    Wien's Displacement Law 35
    Planck's Law 36
    Wien's Approximation Law 37
    Rayleigh-Jeans Law 38
    Stefan-Boltzmann Law 40
  Inverse Square Principle 41
  Irradiance/Radiant Flux Density 42
  Hertzsprung-Russell Diagram 43
  Comments on H-R Diagram 44
  Eddington Luminosity 46
  Sérsic Profile 47
  Transit Method 50

4 Newton's Laws 53
  Newton's Three Laws of Motion 53
  Gravity 54
  Gravitational Potential Energy 57
  Virial Theorem 59
    Jean's Criterion 60
  Escape Velocity 64
  Roche Limit 64
    Tidal Forces 65

5 Kepler's Laws 69
  Ellipses 69
    Ellipticity 73
  Orbital Elements 74
  Kepler's Laws 77
    Orbital Velocity 78
    Orbital Period/Mass 82
    Mass of Sagittarius A* 85
    Geostationary Satellites 89
  Lagrange Points 90
  Deviations from Kepler's Laws Reveal Dark Matter 91

6 Climbing the Distance Ladder 101
  Distance Formula 101
  Parallax 102
  Galaxy Distance via Cepheids 105
    Apparent vs Absolute Magnitude 107
    Period-Luminosity Relation 109
    Distance Modulus 110
  Type 1a Supernovae 113
  Tip of the Red Giant Branch 114
  Tully-Fisher Relation 115
  Galaxy Diameter 118

Contents   xiii

7 Hubble's Law of the Universe 121
   Doppler Effect/Redshift 121
   21-cm Neutral Hydrogen 126
   Hubble Parameter/Constant 127
   Lookback Time 129
   Expansion Scale Factor 130

8 Relativity 133
   Special Relativity 133
   Two Postulates of Relativity 135
   Time Dilation 136
   Lorentz Dilation Factor 140
   General Relativity 141
   Time Dilation 142
   Global Positioning System 144
   Apsidal Precession 145
   Gravitational Redshift 148

9 Black Holes 161
   Schwarzschild Radius 162
   M87's Black Hole 164
   Central Velocity Dispersion 166
   Black Hole Density 168
   Schwarzschild Precession for Black Holes 169
   Hawking Radiation/Temperature 170
   Black Hole Luminosity 172
   Black Hole Evaporation 173
   Black Hole Entropy 173

10 The Universe 175
   Einstein Field Equations (EFE) 175
   Geometry of Space 177
   First Friedmann Equation 178
   Eras Tour 180
   Second Friedmann Equation 183
   Critical Density 185
   Density Parameter 186
   Space is Flat 187
   Spacetime Metrics 189
   Minkowski Metric 192
   Schwarzschild Metric 192
   De Sitter Metric 194
   Friedmann-Lemaître-Robertson-Walker Metric 195
   Topology of the Universe 196
   Future of the Universe 197
   Equation of State Parameter 198

| 11 | **Epilogue: Are We Alone?** | 201 |
| | Drake Equation | 201 |

Appendixes 205

Terminology 229

Further Reading 243

Index 245

Chapter 1
Measurement

> I often say that when you can measure what you are speaking about, and express it in numbers, you know something about it; but when you cannot measure it, when you cannot express it in numbers, your knowledge is of a meagre and unsatisfactory kind
>
> Lord Kelvin

Scientific Notation

When dealing with astronomical quantities such as distance, the numbers involved are indeed truly astronomical. Or in some cases, infinitesimally small. So, we need appropriate notation in dealing with them. The first simplification is to use powers of 10 notation, so that we write 6,000,000 as 6×10^6, and 6/1,000,000 as $6 \times 10^{-6} = 0.000006$. But of course, we need units to go with these quantities and both the large and small quantities come from the nature of light and other electromagnetic radiation. In some instances, if there are not too many zeros to contend with, the number itself can be written without the powers of 10 notation. Scientific measurements are invariably done in the metric system which is very amenable to scientific notation as will become evident.

Example Suppose we multiply

$$4321 \times 123 = 531{,}483.$$

Doing this in powers of 10,

$$(4.321 \times 10^3) \times (1.23 \times 10^2) = 5.31483 \times 10^5.$$

Here we are faced with a value involving three decimal places times one involving two decimal places and ending up with an answer involving five decimal places. We

© The Author(s), under exclusive license to Springer Nature Switzerland AG 2025
J. L. Schiff, *Basic Mathematical Insights into Astrophysics*, Undergraduate Lecture Notes in Physics, https://doi.org/10.1007/978-3-031-79100-0_1

will encounter this situation often and at least in this instance, both values 5.31×10^5, and 5.315×10^5, are suitable rounded-off answers. The general practice is to round off to the number with the least number of significant figures, in this case three. However, *we will not worry too much about the number of decimal places in our answer as long as it looks reasonable just to keep life simple.* As we will encounter, there are differing values of celestial distances and other important quantities depending on the method employed. Indeed, very often averages are taken of a set of differing values obtained by various techniques.

SI Units

The International System of Units is the main standard scientific system of units. The units that appear in the sequel are:

Time: second (s)
Length: meter (m)
Mass: kilogram (kg)
Temperature: Kelvin (K)
Force: Newton (N) = kg · m/s^2
  The units of force derive from the formula $F = ma$ (see Chap. 4) which has units kg · (m/s^2).
Energy: Joule (J) = N · m = kg · m^2/s^2
  The units of energy derive from the work done by a force of 1 Newton in moving a mass a distance of 1 m (in the direction of the force).
Power: Watt (W) = 1 J/s

The famous physicist Max Planck introduced a new set of units with the view that they would be independent of all cultures and would even be meaningful to extraterrestrials (!) as they only depend on universal constants (in our Universe of course):

Planck Length: $l_P = \sqrt{\frac{\hbar G}{c^3}} = 1.616\ 255 \times 10^{-35}$ m;

Planck Time: $t_P = \frac{l_P}{c} = \sqrt{\frac{\hbar G}{c^5}} = 5.391\ 247 \times 10^{-44}$ s;

  (This is the time it takes for a photon of light to travel a distance of Planck length).

Planck Mass: $m_P = \sqrt{\frac{\hbar c}{G}} = 2.176\ 434 \times 10^{-8}$ kg;

Planck Temperature: $T_P = \sqrt{\frac{\hbar c^5}{G k_B^2}} = 1.416\ 784 \times 10^{32}$ K.

Constants

The International Astronomical Union has adopted various values to be standard constants so that astronomers are all working with the same data. Often the values are rounded off for convenience of calculation. The units are the from the International System of Units (SI) used by scientists the world over. Here are the ones that will appear in this text in all their glory:

Gravitational Constant: $G = 6.67430 \times 10^{-11}$ m^3/kg · s^2
Earth's Gravitational Acceleration: $g = 9.80665$ m/s^2
Mass of Earth: M_E ($M_\oplus$) $= 5.9722 \times 10^{24}$ kg
Radius of Earth: R_E ($R_\oplus$) $= 6371$ km[1]
Mass of Moon: $M_M = 7.346 \times 10^{22}$ kg
Radius of Moon: $R_M = 1.737 \times 10^3$ km
Apparent Magnitude of Moon: -12.74
Mass of Sun: $M_\odot = 1.9885 \times 10^{30}$ kg
Radius of Sun: $R_\odot = 696{,}340$ km
Absolute Magnitude of Sun: $M_\odot = +4.83$.[2]
Apparent Magnitude of Sun: -26.74
Surface Temperature of Sun: $T_\odot = 5772$ K
Luminosity of Sun: $L_\odot = 3.837 \times 10^{26}$ W
Length of (Julian) Year: 365.25 days
Speed of Light in a Vacuum: $c = 299{,}792{,}458$ m/s
1 Light-year: 9,460,730,472,580,800 m
1 Astronomical Unit (AU): 149,597,870,700 m $= 1.58125 \times 10^{-5}$ ly
1 Parsec: 3.2616 light-years $= 3.0857 \times 10^{16}$ m
Sidereal Year: 365.25636 days (the time taken for the Earth to complete one orbit of the Sun relative to the fixed stars)
Sidereal Day: 23 h, 56 min, 4.0905 s (the time taken for the Earth to complete one rotation relative to the fixed stars)
Julian Year: 365.25 days (average length of a year in the Julian calendar)
Julian Day: 86,400 s
Planck Constant: $h = 6.626\,070\,15 \times 10^{-34}$ m^2 · kg/s
Reduced Planck Constant: $\hbar = 1.054\,571\,818 \times 10^{-34}$ m^2 · kg/s
Boltzmann Constant: $k_B = 1.380\,649 \times 10^{-23}$ J/K
Stefan-Boltzmann Constant: $\sigma = \frac{2\pi^5 k^4}{15 c^2 h^3} = 5.670\,374 \times 10^{-8}$ W · m^{-2} · K^{-4}.

---

[1] This is the *authalic radius*, that is, the radius of a sphere having the same surface area as the Earth. This is a good compromise radius as the Earth is not a perfect sphere but rather an oblate spheroid.

[2] The same symbol $M_\odot$ is commonly used to indicate both the mass of the Sun as well as its absolute magnitude so the context is important to distinguish the two. Note the Sun symbol ⊙ which is standard.

Remark At this juncture we should mention that all constants are rounded off at some stage and so the values are actually only approximations up to a certain number of decimal places as determined by the available evidence. Therefore, all calculations using these constants are going to be approximations as well. Just like writing down any value of the constant π is going to be an approximation because the decimal expansion of π is never ending. But in order to do calculations with π or any of the above constants, we utilize as many decimal places as is reasonable for that particular calculation. And in our calculations, we write an *equal sign* which is the appropriate thing to do, although on some occasions we will use the symbol $\approx$ for a more general approximation. Astronomy is considered a natural science like Geology or Biology since we are dealing with the real world and we need to bear that in mind.

Example Taking the length of a Julian year as 365.25 days, since there are 86,400 s per day $\left(60 \frac{s}{min} \times 60 \frac{min}{h} \times 24 \frac{h}{day}\right)$. A figure that will be of use in the sequel is the following:

No. of seconds per year $= 86{,}400 \text{ s/day} \times 365.25 \text{ days} = \mathbf{31{,}557{,}600 \text{ s/year}}$.

This value will do for your purposes. On the other hand, the number of seconds in a sidereal year is **31,558,149.5 s**. If we round off both results suitably, say to **3.156×10^7 s/year**, we need not make any distinction in our calculations.

What's the Angle?

It is useful to use both degrees and radians in astronomical calculations. A *degree* (*deg*) is the plane angle subtended by 1/360th of the full circle and denoted by the symbol °. We need fractions of a degree, and so 1/60th of a degree is called an *arcminute* (*arcmin*), denoted by the symbol ′, and 1/60th of an arcminute is an *arcsecond* (*arcsec*) denoted by ″. Thus, there are 3,600 arcseconds in a degree. These units follow the same sexagesimal system as do units of time.

Now if we take a circle of radius equal to r, then what we mean by *radian* (*rad*) is the angle that is subtended by the length of the radius placed along the circumference of the circle. So, length r along the circumference corresponds to 1 radian which means that the whole circumference of length $2\pi r$ corresponds to 2π radians (Fig. 1.1). In other words, 360° corresponds to 2π radians. Now we have a means to convert degrees into radians (and vice-versa).

Thus, we have the ratios

$$\frac{2\pi \text{ (rad)}}{360°} = \frac{\pi \text{ (rad)}}{180°} = \frac{x \text{ (rad)}}{y \text{ (deg)}},$$

Fig. 1.1 The angle θ is equal to 1 radian (rad). The size of the circle is immaterial. (Courtesy Katy Metcalf)

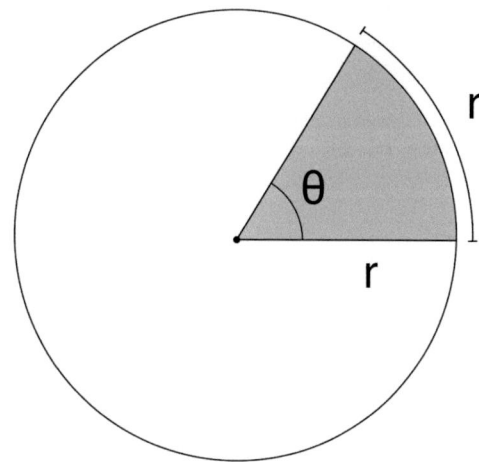

where one of x or y are known and you want to find the other. For example, if $y = 45°$, then $x = 45° \times \frac{\pi (rad)}{180°} = \frac{\pi}{4}$ rad. In general,

$$x \text{ (rad)} = y \text{ (deg)} \times \frac{\pi \text{ (rad)}}{180°}.$$

On the other hand, to convert $\frac{\pi}{3}$ rad to degrees, the above ratios imply that $y \text{ (deg)} = \frac{\pi}{3} \text{ rad} \times \frac{180°}{\pi \text{ rad}} = 60°$. More generally,

$$y \text{ (deg)} = x \text{ (rad)} \times \frac{180°}{\pi(\text{rad})}.$$

What will be useful in the discussion of a parsec in the sequel is the conversion of 1 arcsec ($=1/3{,}600°$) to radians which gives

$$1 \text{ arc sec} = \frac{1}{3600} \times \frac{\pi}{180} = \frac{\pi}{648{,}000} \text{ rad} = 4.848 \times 10^{-6} \text{ rad}. \quad (1.1)$$

In view of the large distances involved, one common unit of angle is the *milli-arcsecond* which is a thousandth of an arcsecond

$$1 \text{ mas} = 0.001 \text{ arc sec} = 10^{-3} \times 4.848 \times 10^{-6} = 4.848 \times 10^{-9} \text{ rad}, \quad (1.2)$$

and lastly the *micro-arcsecond*, denoted by μ as, and $1 \, \mu$ as $= 10^{-6}$ arcsec so that

$$1 \, \mu\text{as} = 10^{-6} \times 4.848 \times 10^{-6} = 4.848 \times 10^{-12} \text{ rad}. \quad (1.3)$$

Fig. 1.2 Illustration of a circle of radius r with a segment of the circumference having length S which spans an angle θ. Courtesy Katy Metcalf

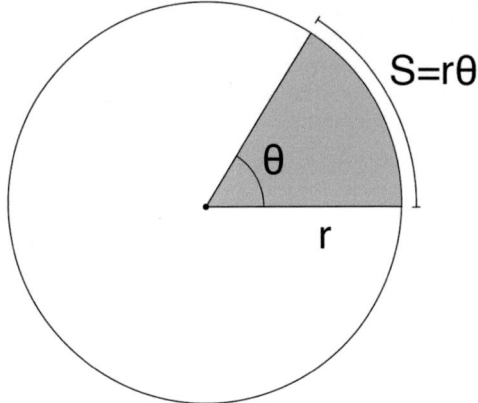

Note that we have dropped the units in the calculation for simplification purposes and the reader can check that the units are correct. This will be convenient to do from time to time.

Exercise
(a) Determine how many degrees equal 1 radian. *Ans.* 57.296.
(b) From (a) determine how many arcseconds equal 1 radian. *Ans.* 206,265 arcsec per radian.

Exercise
Convert 15 μas to radians. *Ans.* 72.72×10^{-12} rad.

From the above equations we can derive a very elementary formula. On a circle of radius r, if S represents the length of an arc of a circular sector which spans an angle of θ radians with the origin, then we have the relation

$$\frac{S}{2\pi r} = \frac{\theta}{2\pi},$$

and this implies that

$$\boxed{S = r\theta}. \tag{1.4}$$

See Fig. 1.2. This formula will re-appear in Chap. 6 in the discussion of galaxy diameters and in other contexts.

Solid Angles

Analogously, there are also solid angles that appear in Astronomy. A *steradian* (sr) is the solid angle formed by the origin of a sphere of radius r that subtends a circular area on the sphere of $A = r^2$ (Fig. 1.3).

Fig. 1.3 A steradian formed by the origin and a circular area on the sphere of radius r. (Courtesy Katy Metcalf)

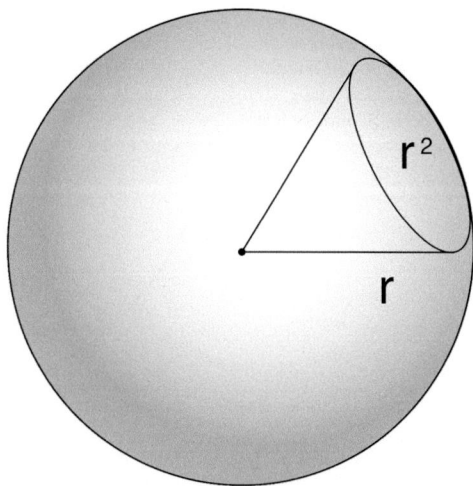

Since the sphere of radius r has a surface area equal to $S = 4\pi r^2$, this means that the sphere contains 4π steradians just as the circle contains 2π radians.

In general, a spherical cap of area A forms a solid angle Ω in the sphere of radius r by the relation

$$\frac{A}{r^2} = \frac{\Omega}{1 sr}$$

which is to say that,

$$\boxed{\Omega = \frac{A}{r^2} \, sr}. \tag{1.5}$$

We can say something further since from Geometry as we know that a spherical cap has an area determined by the formula

$$A = 2\pi r h,$$

where r is the radius of the sphere and h is the height of the cap as in Fig. 1.4.

Moreover, there is the obvious trigonometric relation

$$\frac{r-h}{r} = \cos\theta/2$$

so that solving for h we have

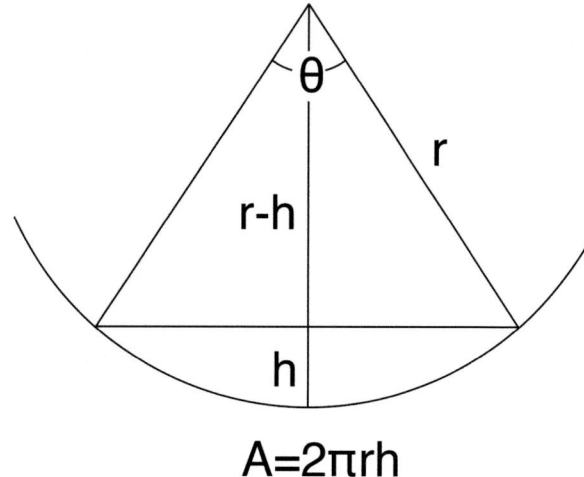

Fig. 1.4 Depiction of spherical cap subtending an angle θ of height h and having surface area $A = 2\pi r h$. (Courtesy Katy Metcalf)

$$h = r(1 - \cos\theta/2).$$

Thus, we can write steradians in terms of the subtended angle θ in view of Eq. (1.5) and the preceding

$$\boxed{\Omega = \frac{A}{r^2}\ (sr) = \frac{2\pi h}{r}\ (sr) = 2\pi(1 - \cos\theta/2)\ (sr)}. \qquad (1.6)$$

Note here that the radius of the sphere is no longer needed and only the subtended angle θ is relevant. Steradians are a key component in various aspects of light measurements as discussed in Chap. 3.

Example
The Sun subtends an angle of 0.57° as viewed from Earth. If we assume the Sun to be a circular spot on a sphere, then we can calculate the steradians formed by the Sun. First, we convert θ to radians,

$$\theta = 0.57 \times \pi/180 = 0.00995 \text{ rad},$$

which gives

$$\Omega = 2\pi(1 - \cos\theta/2) = 7.79 \times 10^{-5} \text{ sr}.$$

Exercise
The full Moon subtends an angle of 30 arcmin as viewed from Earth. What solid angle does it form? *Ans.* 5.9×10^{-5} sr.

Light-Years/Parsecs

A convenient unit of measurement when dealing with distance within our own Solar System is the *Astronomical Unit* (*AU*) that represents the average distance from the Earth to the Sun.[3] This is equivalent to 149.6 million km (or more precisely 149,597,870,700 m) and for example, the distance of the Asteroid Belt from the Sun (that lies between Mars and Jupiter) is 2.2 to 3.2 AU and the distance to dwarf planet Pluto is 39 AU. But if we want to get to the stars and explore the vastness of the Universe, we require units of measurement that are better suited to the task. We will use two units, one in common parlance and one used by professional astronomers. The former is the *light-year* (*ly*) which is the distance light travels in 1 year. This is easy enough to calculate given that in the preceding we have determined that there are $=3.156 \times 10^7$ s/yr, and at a velocity of $v = c = 2.998 \times 10^5$ km/s we use the elementary relation that the *distance equals the velocity times time*,

$$\boxed{D = v \cdot t}$$

$$= \left(2.998 \times 10^5 \frac{\text{km}}{\text{s}}\right) \times \left(3.156 \times 10^7 \frac{\text{s}}{\text{yr}}\right)$$

$$= 9.462 \times 10^{12} \text{ km in one year,}$$

that is, 1 ly *equals roughly 9.5 trillion km* (nearly 6 trillion miles). Even with this enormous distance many galaxies will be millions, even billions, of light-years distant from the Earth. The closest star to the Earth is Proxima Centauri at a distance of 4.2 ly and the nearby galaxy, Andromeda, is at a distance of 2.5 million ly.

Exercise How many astronomical units in 1 light-year? *Ans.* ≈63,241 AU/ly.

Exercise Pulsar PSR J1748-2446ad, the fastest spinning pulsar known (as of 2024), rotates at the rate of 716 revolutions per second. It has a radius of <16 km. How fast is a point on the equator rotating assuming a 16 km radius? *Ans.* ≈72,000 km/s ≈ 24% the speed of light.

One issue that immediately arises is that if we say that a galaxy is 10 million light-years away, that is, the light from the galaxy has taken 10 million years to reach us, then due to the fact that the Universe is continually expanding (which will be dealt with in Chap. 7), the galaxy is no longer 10 million light-years distant. It is actually much further and moreover, the galaxy in the meantime may have collided with another galaxy or significantly changed its appearance. Thus, a galaxy that has been measured to be at a distance 10 million light-years is being seen by us on Earth as it was 10 million years ago, and we have looked into the distant past. Nevertheless, we

---

[3]The orbit of the Earth about the Sun is very slightly elliptical as will be discussed in Chap. 5.

will consider such a galaxy as being viewed in the *here and now* in order to avoid any confusion as to what may have happened to the galaxy in the intervening 10 million years.

The other unit of measurement favored by astronomers is the parsec that will be discussed further in the section on parallax in Chap. 6 and is equal to 3.2616 light-years. So, there is not much difference between light-years and parsecs, and we will use both interchangeably.

Exercise Convert 1 parsec to meters. *Ans.* 3.0857×10^{16} m.

Mass

The mass of stars, galaxies and other objects in the Universe will be measured in terms of the mass of our Sun, namely *solar masses* ($M_\odot$) which is roughly 2×10^{30} kg. Thus, the mass of the red giant star Betelgeuse is 16.5 – 19 $M_\odot$[4] and that of our Milky Way galaxy has been estimated to be $\sim 10^{12}$ solar masses.[5]

Density

We will consider most of our round celestial objects as perfect spheres which of course they are not quite due to rotation which makes them slightly oblate. The formula for the volume V of a sphere of radius R is

$$V = \frac{4}{3}\pi R^3.$$

To compute the average density ρ of our spherical object we divide its mass M by its volume, that is

$$\rho = \frac{M}{V}.$$

As an example, let us consider the Sun which has a mass $M = 1.99847 \times 10^{30}$ kg and a radius of $R = 6.9634 \times 10^5$ km. Then the volume will be

---

[4] M. Joyce et al., Standing on the Shoulders of Giants: New Mass and Distance Estimates for Betelgeuse through Combined Evolutionary, Asteroseismic, and Hydrodynamic Simulations with MESA, *ApJ*, 902 (2020), 25 pp.

[5] G. Fragione, A. Loeb, Constraining the Milky Way mass with hypervelocity stars, *New Astronomy*, 55 (2017), 32–38.

$$V = \frac{4}{3}\pi(6.9634 \times 10^5 \text{ km})^3$$
$$= 1.41433 \times 10^{18} \text{ km}^3.$$

So much for the volume. Now the average density is

$$\rho = \frac{M}{V} = \frac{1.99847 \times 10^{30} \text{ kg}}{1.41433 \times 10^{18} \text{ km}^3}$$
$$= 1.413 \times 10^{12} \frac{\text{kg}}{\text{km}^3}.$$

This value is somewhat meaningless so let us convert it to something more familiar. Thus

$$\rho = 1.413 \times 10^{12} \frac{\text{kg}}{\text{km}^3} \times \frac{10^3 \text{gm}}{\text{kg}} \times \frac{\text{km}^3}{10^9 \text{m}^3} \times \frac{\text{m}^3}{10^6 \text{cm}^3}$$
$$= 1.413 \frac{\text{gm}}{\text{cm}^3},$$

which is slightly denser than that of water (1 gm/cm^3)!

Exercise Determine the mean density of the Earth in gm/cm^3. *Ans.* 5.51 gm/cm^3.

Temperature

Everything in the Universe has a temperature. Even 'empty space' has a temperature due to the Cosmic Background Radiation (CMB) left over from the Big Bang. The unit is a *Kelvin* (K) and the relationship to the more familiar degrees Celsius is

$$\textbf{Kelvin} = \, ^\circ\textbf{C} + \textbf{273}.$$

The lowest theoretically possible temperature is 0 K and therefore equal to -273 °C[6] and is known as *absolute zero*. The temperature of the CMB is 2.7 K which is the temperature of empty space and nothing is known to be at absolute zero, either in space or in a laboratory.

Most temperatures that we encounter will be in millions of K and since the difference with °C is only 273 °C we may consider them as essentially the same.

---

[6] More precisely, -273.15 °C.

Logarithms

Logarithms are very useful in working with large quantities as Astronomy is replete with very large numbers. We will almost always use base 10 and recall that a logarithm is the *exponent* of 10 that is equal to the number in question. So, the logarithm of 1,000,000,000 is 9 since $10^9 = 1,000,000,000$. On the other hand, if you have the value of the logarithm, say 6, then the number itself is $10^6 = 1,000,000$. In general the numbers are not so trivial and one has to use a calculator to work out that the logarithm of 1234, written $\log_{10} 1234$, equals 3.0913. Or if $\log_{10} M = 9.81$ then

$$M = 10^{9.81} = 6.46 \times 10^9.$$

Brightness/Stellar Magnitudes

Some stars, like our Sun and Sirius obviously appear brighter than others, while some stars are very faint, requiring dark skies and good eyesight, or a telescope to be seen. We are using the general term 'brightness' to mean the perceived intensity of light emanating from a source[7] and we all know when we see one star as brighter than another. Quantifying the brightness differences denoted by stellar magnitudes is based on an old notion, namely that the brightest star which we say has magnitude $m = 1$ and the faintest star that can be seen by the naked eye has magnitude $m = 6$. This gives rise[8] to a logarithmic scale by which the star with $m = 1$ is considered to be 100 times brighter than a star of $m = 6$. What counts here is the *difference* between the two magnitudes so that *a difference in stellar magnitudes of 5 means that one star is 100 times brighter than the dimmer one.*

Unwinding this basic principle, if we start out with a star of brightness magnitude b, then the next magnitude will be kb where k is the multiplication factor of the change in magnitude. The next stellar magnitude will be denoted by $k^2 b$, followed by $k^3 b$, $k^4 b$, and finally the sixth stellar magnitude will be $k^5 b$. If this latter is to be $100b$, then we have

$$k^5 b = 100 b,$$

so that the multiplication factor is

---

[7]There are two types of stellar brightness, *apparent* (as that viewed from Earth) and *intrinsic* (emitted light, independent of distance) and these are discussed further in Chap. 6.

[8]Introduced by English astronomer N.R. Pogson in 1856 although the idea of a 6-tier scale goes back to the Greek astronomer Ptolemy.

$$\boxed{k = \sqrt[5]{100} = 2.512\ldots},$$

which means that each stellar magnitude varies from the one preceding/succeeding it by roughly 2.5 times. The reader might come up with a simpler system but this is what is used in Astronomy so best to go along with it.

Just to keep everyone alert, the magnitude scale is arranged so that *higher magnitude numbers indicate lower brightness*, or, *lower magnitude numbers indicate higher brightness*, so that a magnitude 2 star is ~2.5 times brighter than a magnitude 3 star and is ~6.25 times brighter than a magnitude 4 star. The bright star Vega is of magnitude $m = 0.03$ and an even brighter star is Sirius of magnitude $m = -1.46$. Thus we are free to use negative values of magnitude but the same rule applies, i.e., a magnitude $m = -1$ star is ~2.5 times brighter than a magnitude $m = 0$ star. It is the difference in magnitudes that determines the brighter/dimmer factor so that a difference of say 4 magnitudes between two stars means that the brighter star is a factor of $2.5^4 \approx 40$ times brighter than the dimmer star.

Describing this scenario in mathematical terms, suppose that two stars are of brightness b_1 and b_2 respectively and that $b_1 < b_2$. However, we do know that lower magnitude stars are brighter than those of greater magnitude, so that $m_1 > m_2$. The *factor* by how much brighter b_2 is than b_1 depends as we have seen on the difference of their respective magnitudes $m_1 - m_2$ in the form

$$b_2 = b_1 \times 2.512^{(m_1 - m_2)},$$

and since $2.512 = \sqrt[5]{100}$, it follows that

$$\boxed{\frac{b_2}{b_1} = 100^{\frac{(m_1 - m_2)}{5}}}. \tag{1.7}$$

Thus, for example, the apparent magnitude of the night sky's bright star Sirius is -1.46 and that of Canopus the second brightest star is -0.74. Therefore, if b_1 is the brightness of Canopus, and b_2 is the brightness of Sirius so that $b_1 < b_2$, then letting $m_1 = -0.74$ (Canopus) and $m_2 = -1.46$ (Sirius), we have $m_2 < m_1$ and by Eq. (1.7)

$$\frac{b_2}{b_1} = 100^{\frac{(-0.74 + 1.46)}{5}}$$

Doing the calculation, we find that

$$\frac{b_2}{b_1} \approx 1.94,$$

which is to say that Sirius is roughly twice as bright Canopus.

Exercise Determine how many times the brighter the Sun is compared to the full Moon where the Sun has an apparent magnitude of -26.74 and the full Moon has an apparent magnitude of -12.74. *Ans.* ~398,000 times brighter.

Galactic Orbital Velocities

A key consideration in the study of spiral galaxies such as our own Milky Way, is the orbital velocities of its stars and gas about its center, usually measured in km/s. For example, the Sun orbits the center of the Milky Way at an orbital velocity of about 220 km/s. In the context of a spiral galaxy, the orbital velocity of a star will be denoted by, v_{ORB}.

But as will be seen in Chap. 7, all that can be measured in other galaxies is the *line-of-sight velocity*, v_{LOS}, so that a little geometry comes into play since most spiral galaxies will not be edge on. As well, the v_{LOS} value includes that of the galaxy's rotation as well as the velocity of the expanding Universe. However, the latter is essentially negligible in the context of the rotational velocities of stars in spiral galaxies so we can safely ignore it.

Thus, we need to take into consideration the angle of tilt of the galaxy to the plane of the sky. This angle between the plane of the sky and the plane of the galaxy is the *inclination, i*, and can vary between 0° and 90°, so that when the galaxy is *face on* the inclination is $i = 0°$ and when the galaxy is *edge on* its inclination is $i = 90°$.

The line of sight (LOS) along which we view the galaxy from Earth is the line perpendicular to the plane of the galaxy. In the face on case, the orbital velocity, v_{ORB}, of particular stars/gas cannot be determined as they are perpendicular to our line of sight. However, in the edge on case, we have $v_{ORB} = v_{LOS}$.

In general, taking into account the inclination of the galaxy, the vector component of the rotational velocity in the direction of the line of sight is determined by the geometry of Fig. 1.5:

$$\boxed{v_{LOS} = v_{ORB} \cdot \sin i}. \tag{1.8}$$

The important thing to note is that the velocity vector of a star in the galaxy lies in the plane of the galaxy. Moreover, when the galaxy is face on and $i = 0°$, then $\sin 0 = 0$ which means that we cannot determine the rotational velocity, and when the galaxy is edge on, $i = 90°$ and $\sin 90° = 1$ so that $v_{LOS} = v_{ORB}$.

Orbital velocities will come into their own in the study of Kepler's Laws in Chap. 5 since many celestial objects are in orbit about something else, such as a planet, star, galactic center, black hole etc.

Fig. 1.5 An oblique view of a star in a spiral galaxy with inclination i with respect to the plane of the sky with the rotational velocity vector of the star lying in the galactic plane. (Courtesy Katy Metcalf)

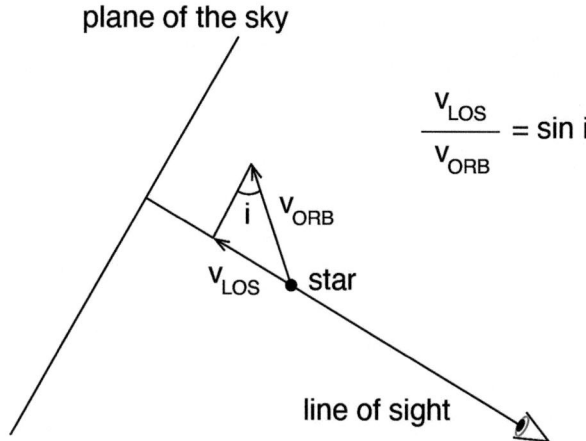

Method of Least Squares

This is a very useful method that will appear in the sequel when we want to fit a straight line to a set of data. Astronomers gather data points in two dimensions with (x, y) coordinates and often this data appears to lie on a straight line exhibiting a linear relation between y and x. But what straight line (called a *regression line*) best represents the data? This is an issue in various scientific studies where data is accumulated, not just Astronomy.

To this end, let us consider a set of n data points $(x_1, y_1), (x_2, y_2), \ldots, (x_n, y_n)$. We know that a straight line is represented by the formula $y = ax + b$. The basic idea is quite simple in that the regression line is the straight line that minimizes the sum of the discrepancies d_i between the data and the line (Fig. 1.6). This represents the total error S between the line and the data but some of the data points are above the regression line and some will be below. As we want the actual vertical distance between the points and the regression line, we eliminate the distinction between points above and below the line by taking the square of each value d_i.

Then we sum their squares to get a measure on the total amount of discrepancy, i.e.,

$$E = \sum_{i=1}^{n} d_i^2,$$

where $d_i^2 = (ax_i + b - y_i)^2$, for each $i = 1, 2, 3, \ldots, n$. The quantity E is the one we wish to minimize and thus it becomes a problem in Calculus.

The details of the derivation of the regression line are left to Appendix 1, where it is shown that the error quantity E is minimized when

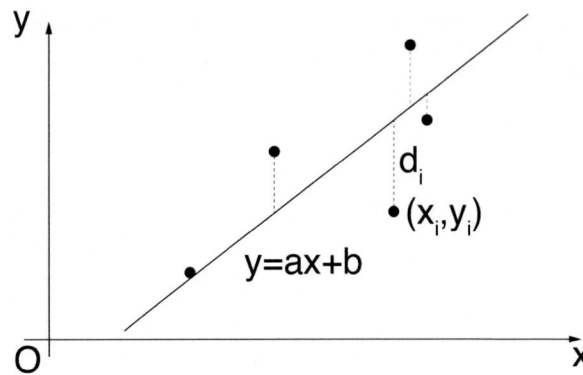

Fig. 1.6 In the method of least squares, the 'best' straight-line fit minimizes the vertical distance between the line and the data points. (Courtesy Katy Metcalf)

$$a = \frac{n\sum_{i=1}^{n} x_i y_i - \left(\sum_{i=1}^{n} x_i\right)\left(\sum_{i=1}^{n} y_i\right)}{n\sum_{i=1}^{n} x_i^2 - \left(\sum_{i=1}^{n} x_i\right)^2}, \quad b = \frac{\sum_{i=1}^{n} y_i - a\sum_{i=1}^{n} x_i}{n}. \quad (1.9)$$

Although the formulas look complicated, they simply represent sums and products of the coordinate points (x_i, y_i) and as such represent nothing more than a lot of arithmetic.

Example Let us find the regression line for the data set of points: (1, 6), (2, 6), (3, 8), (4, 12).

Therefore,

$$a = \frac{4[(1)(6) + (2)(6) + (3)(8) + (4)(12)] - [(1 + 2 + 3 + 4)(6 + 6 + 8 + 12)]}{4\left[1^2 + 2^2 + 3^2 + 4^2\right] - (1 + 2 + 3 + 4)^2}$$

$$= \frac{360 - 320}{120 - 100} = 2.$$

Then,

$$b = [(6 + 6 + 8 + 12) - 2(1 + 2 + 3 + 4)]/4$$

$$= \frac{32 - 20}{4} = 3.$$

We conclude that the best fit straight line is given by $y = 2x + 3$.

Since the data involved in most real-life calculations is generally not so accommodating as in this example, fortunately there are applications in most statistical packages such as 'R' or 'SAS', as well as online sources[9], that do the linear regression calculations for you.

---

[9] For example: https://www.statskingdom.com/linear-regression-calculator.html

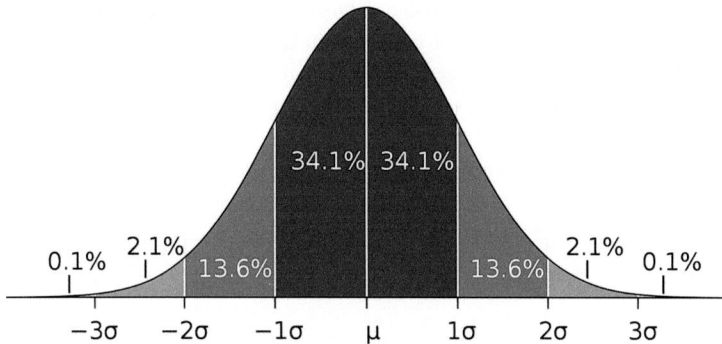

Fig. 1.7 In a normal distribution curve defined by Eq. ($_{1.10}$) with mean value μ and standard deviation σ. In any such a distribution, ~68% of the values will cluster about the mean within 1 standard deviation written as 1σ. Likewise, ~95% of the data will be within two standard deviations (2σ) of the mean, and ~99.7% of the data will be within 3σ of the mean. Image Public Domain

Exercise Find the regression line that best fits the data points: $(1, 1)$, $(2, 6)$, $(3, 7)$, $(4, 12)$. Ans. $y = 3.4x - 2$.

Exercise Find the regression line that best fits the data points: $(1, 2)$, $(3, 5)$, $(5, 7)$, $(7, 10)$, $(9, 15)$. Ans. $y = 1.55x + 0.05$.

Standard Deviation

A lot of astronomical data involves the use of Statistics. We will only touch upon the most basic elements which are sufficient to understand much of what occurs in the analysis of the data. One very common occurrence is when there are few instances of a property taking extremely high or extremely low values with the majority of values not too far removed from some average, forming a bell-shaped curve. This known as a normal (gaussian) distribution and has the shape indicated in Fig. 1.7.

It is often convenient in our graphical depiction to take the average (mean) value to be at the origin so that the curve is symmetric about $x = 0$. An equation describing this curve is given by an exponential function with a negative power,

$$y = \frac{1}{\sqrt{2\pi}} e^{-x^2/2}.$$

However, this is merely an idealization since the value of y tends to zero as x becomes increasingly large (or small) and with real data the x coordinate does not increase without bound.

Now let us say that the mean is at some value, customarily denoted by μ.[10] In Fig. 1.7 it is easy to imagine that the central bulge about the value μ to be much narrower or much wider depending on whether much of the data falls near the mean or is more widely dispersed. The manner in which the data is actually dispersed around the mean value is a very important consideration known as the *standard deviation* σ and for a set of data values $y_1, y_2, \ldots y_N$ (from points $x_1, x_2, \ldots x_N$ on the x-axis) and is given by

$$\sigma = \sqrt{\frac{(y_1 - \mu)^2 + (y_2 - \mu)^2 + \ldots (y_N - \mu)^2}{N}} = \sqrt{\sum_{i=1}^{N} \frac{1}{N}(y_i - \mu)^2}.$$

Note that squaring the differences between the data values y_i and the mean μ has the effect of washing out any distinction between positive and negative values as was done in the method of least squares.

Now it is possible to describe the normal distribution curve in terms of a general mean value μ as well as its standard deviation σ, that is

$$\boxed{y = \frac{1}{\sigma\sqrt{2\pi}} e^{-\frac{(x-\mu)^2}{2}}}. \qquad (1.10)$$

Often astronomers will talk about their data being at the 3-sigma level or 5-sigma level etc., which indicates the statistical confidence that a particular signal was not spurious due to random fluctuations or noise. Such might be a particular signal emitted from a distant galaxy that could indicate intelligent life. If the signal is at a 2-sigma level, then the chances that the signal was *not* due to noise or random fluctuations indicates that it lies within 2-sigma from the mean, thus with a likelihood of ~95% (95%confidence level that the result is not spurious). In other words, there is about a 5% chance, or 1 chance in 20 that the signal was not genuine. If the data is at a 5-sigma level, then the chance of the data being a random fluctuation is 1 in 1.75 million taking a two-sided (two-tailed) criterion and **1 in 3.5 million** taking only one side (one-tailed) into account. This latter is the *gold standard* for claiming a new scientific discovery such as for the Higgs boson confirmed in 2013 or perhaps a fifth force in Nature for which there is currently mounting evidence. A one-tailed test is used when the direction of the effect is known in advance, while a two-tailed test does not assume the direction of the effect.

Exercise What is the chance that a 3-sigma signal is due to a random fluctuation (2-sided criterion)? *Ans.* ~0.3 %.

---

[10] This is the Greek letter mu.

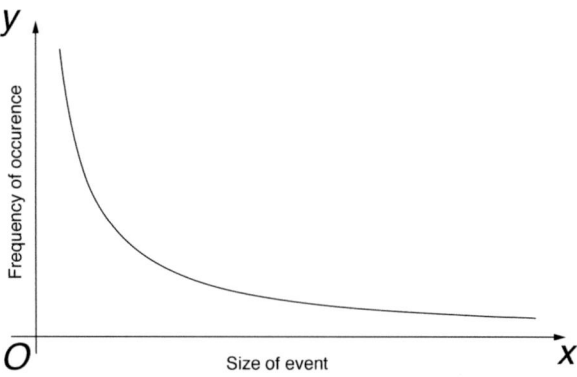

Fig. 1.8 Graph of a power law $y = cx^\alpha$ with $\alpha < 0$. (Courtesy Katy Metcalf)

Power Laws

It is not uncommon for astronomical data in a variety of studies to take the form of a *power law*, that is, one quantity is proportional to a power of some another quantity,

$$y = y(x) \propto x^\alpha.$$

In other words,

$$y = cx^\alpha, \tag{1.11}$$

where c is a constant and α is a constant (positive or negative), and $x > 0$. An example would be the size of asteroid fragments[11] that hit the Earth each year. The vast majority of them are small and increasingly larger fragments are increasingly less likely. A graph with $\alpha < 0$ would look something like Fig. 1.8.

To work with such an equation such as Eq. (1.11), it is convenient to take the logarithm (base 10) of each side, and using common properties of logarithms,

$$\log y = \log c + \alpha \log x.$$

Setting $Y = \log y$, $X = \log x$, and $B = \log c$, we obtain the equation of a straight line,

$$Y = \alpha X + B,$$

where the power α is now the *slope* of the straight line and $B = \log c$ is the value of the Y-intercept. Such a plot of Y vs X is known as a *log-log plot* and facilitates the determination of the values of a and c which can be read off the graph.

---

[11] Most asteroid fragments hitting the Earth come from the Asteroid Belt between Mars and Jupiter. A small number come from the Moon and Mars when they are struck by asteroids.

Power laws also exhibit a notable feature in some astronomical settings and that is one of *scale invariance*. To wit,

$$y(\lambda x) = c(\lambda x)^\alpha = \lambda^\alpha c x^\alpha$$
$$= \lambda^\alpha y(x).$$

This means that the original value of y is simply scaled up by the factor of λ^α to give $y(\lambda x)$, in other words, $y(\lambda x) \propto y(x)$.

Mass-Luminosity Relation

As an example of a power law, stars in a particular mass range on the *main sequence* (see Chap. 3) have an empirically derived mass-luminosity[12] relation given by

$$\boxed{L \approx L_\odot \left(\frac{M}{M_\odot}\right)^{3.5}},$$

where M is the mass of the star, $M_\odot$ the mass of the Sun, L the luminosity of the star, and $L_\odot$ the luminosity of the Sun. This can be written as a log-log relation

$$\boxed{\log\left(\frac{L}{L_\odot}\right) \approx 3.5 \times \log\left(\frac{M}{M_\odot}\right)}, \qquad (1.12)$$

so that the preceding exponent of $a = 3.5$ is now the slope of the straight line going through the origin. A real example from an actual data set is presented in Fig. 1.9.

Exercise Calculate the mass of a star (in solar masses) that has a luminosity of 30 times that of the Sun.

---

[12] The *luminosity*, L, is a measure of the total amount of electromagnetic energy across all wavelengths radiated by the star/galaxy per second and is usually measured in watts (W). The notion appears throughout the text in a variety of settings.

The exponent of 3.5 is valid in the mass range: $2\,M_\odot < M < 55 M_\odot$ whereas the exponent of 4 is valid for stars in the mass range: $0.43\,M_\odot < M < 2M_\odot$, which includes our Sun.

Power Laws 21

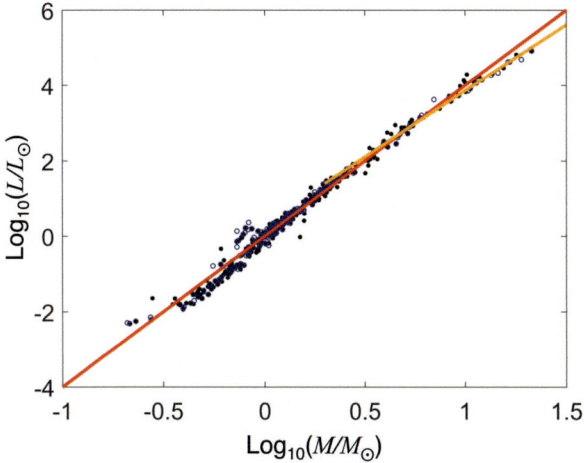

Fig. 1.9 The log-log mass-luminosity relation of a sampling of main sequence stars as discussed in the text, indicating a power law. For a star having mass M, with $55M_\odot > M > 2M_\odot$, then $\log\left(\frac{M}{M_\odot}\right) > \log 2 = 0.3$, the exponent (slope of orange line) is $a = 3.5$. For masses below $2M_\odot$ down to about $\log\left(\frac{M}{M_\odot}\right) > -0.4$, the exponent $a = 4$ is a good fit for the data (red line). Image courtesy J. Wang and Z. Zhong, Revisiting the mass-luminosity relation with an effective temperature modifier, 619, *A&A* (2018), and modified to indicate the orange slope

Main-Sequence Lifetimes

Another example of a power law in astronomy relates to main sequence lifetimes, that is the length of time that a star spends on the main-sequence is governed by its mass according to the approximate relation

$$t_{MS} \approx 10^{10} \times \left(\frac{M}{M_\odot}\right)^{-2.5} \text{ years}, \quad (1.13)$$

where M is the mass of the star and t_{MS} is the lifetime on the main-sequence and 10^{10} years is the approximate lifetime of the Sun on the main sequence, that is for $M = M_\odot$. The exponent of 2.5 is only approximate and can vary somewhat depending on the mass range of the stars in question. It is apparent that the larger the mass, the shorter the main-sequence lifetime as in Fig. 1.10. Eq. (1.13) also means that the Sun's lifetime on the main-sequence is 10^{10} years, that is, 10 billion years and as we will see in the next chapter, roughly half of that time has already been used up.

So, if a star is 10 times more massive than the Sun (approximately, such as Spica A or various B-type stars), then

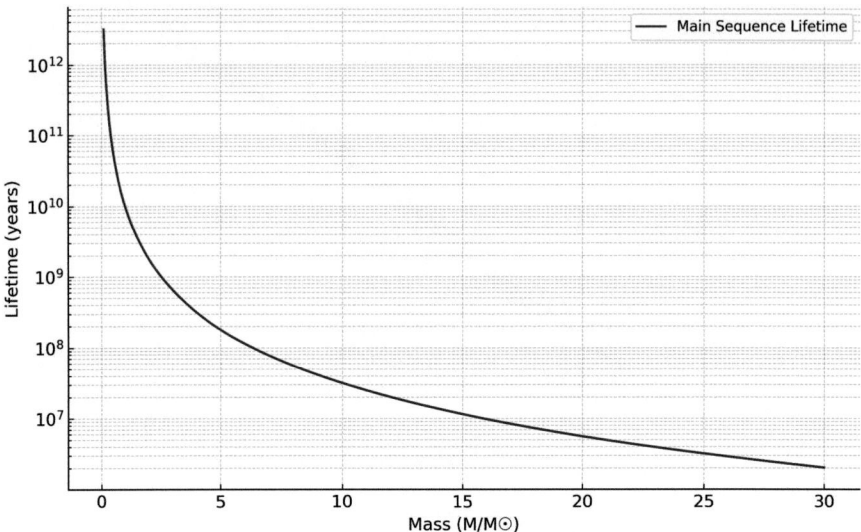

Fig. 1.10 The power law relation $t_{MS} \approx 10^{10} \times \left(\frac{M}{M_\odot}\right)^{-2.5}$ between the mass of a star and its lifetime on the main sequence as discussed in the text. Graph by the author with GPT4

$$t_{MS} = 10^{10} \times 10^{-2.5} = 10^{10} \times 0.00316 \text{ years}$$
$$= \left(3.16 \times 10^{-3}\right) \times 10^{10} \text{ years}$$
$$= 32 \text{ million years.}$$

On the other hand, the red dwarf Proxima Centauri (the closest star to the Sun) is only $0.12 M_\odot$ which gives

$$t_{MS} = 10^{10} \times 0.12^{-2.5}$$
$$= 2005 \text{ billion years.}$$

Chapter 2
Down to Earth

Before venturing into outer space, let us look at our own Earth and see if we can determine some of its basic features, such as its size, age, and mass. The question of the Earth's size,—specifically, its circumference—was a question posed by the Greek mathematician Eratosthenes of Cyrene over 2200 years ago when knowledge of the world was extremely limited. The main mathematical tool he used was some basic geometry espoused by Euclid some years earlier.

Size of Earth

It was noted by Eratosthenes that in the city of Syene (now Aswan, Egypt) that when the Sun was directly overhead at noon during the summer solstice (around the present-day June 21st), the Sun's rays would shine directly down a well. However, in the city where he lived, Alexandria, he noticed that the gnomon of a sundial cast a shadow of 1/50th of a circle ($=7.2°$) as indicated in Fig. 2.1.

The conclusion is that the distance between Syene and Alexandria is 1/50th of the Earth's circumference. Eratosthenes knew this distance to be 5000 *stadia* (plural for stadium) which equates to about 800 km and therefore the Earth's circumference is $50 \times 800 = 40{,}000$ km. The modern value is 40,075 km. Bravo Eratosthenes! And now you know why you studied Geometry in high school and why there is a lunar crater named after him.

Age of the Earth

Astronomy involves ages, often very old ages. For us, one of the most significant old ages is that of our Solar System which formed out of a swirling cloud of gas and dust, which as it cooled, formed the Sun along with all the planets. Again, with some

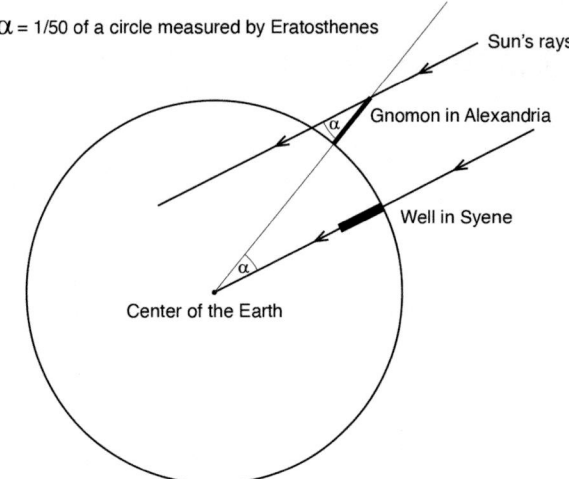

Fig. 2.1 At the time of the summer solstice, an angle α is made with a gnomon in Alexandria, and with the geometry given in the figure, the two angles α are equal opposite interior angles made by a straight line crossing two parallel lines. Thus, the distance between Syene and Alexandria must be $1/50^{\text{th}}$ the circumference of the Earth. Not to scale. (Courtesy Katy Metcalf)

ingenious mathematics we can determine the age of the Solar System to a high degree of accuracy. This is all a consequence of radioactivity.

First, let us explore the equation that describes exponential growth and decay. Some insect populations grow at a rate that is proportional to the number of insects present at that time. More insects, the faster the population grows. And similarly, radioactive substances decay at a rate that is proportional to the amount of the substance that is present at the time. More radium, the faster it decays. This is actually a very curious phenomenon since an atom of radium, which either decays or holds on a bit longer, does not really *know* how much radium is present. Whereas with insects, it makes sense that the more there are, the more opportunities that exist for reproductive purposes.

Both situations lead to a single formula that describes the behavior of their respective rates of growth and decay. Let $N(t)$ be the size of a population or the amount of a radioactive substance at time $t \geq 0$ and let $N(0) \geq 0$ be the population size/quantity at time $t = 0$.

Then the formula for *exponential growth and decay* is given by

$$N(t) = N(0)e^{kt} \qquad (2.1)$$

where k is a positive constant in the case of population growth and a negative constant in the case of radioactive decay. This constant is specific to the population/substance in question. The derivation of the formula is found in Appendix 2.

Since we will be dealing with radioactive substances, we have the notion of *half-life*, denoted by $t_{1/2}$, which is the time it takes for half of the substance to decay. For example, carbon-11 has a half-life of 20.38 min,[1] iodine-131 has a half-life of 8.0197 days, and rubidium-87 (^{87}Rb) has a half-life of 49.15 billion years.[2] The former substance would not be useful for dating the age of the Solar System, but the latter is.

If we know the half-life $t_{1/2}$ of a substance then we can work out its decay constant k. This follows from the expression

$$\frac{N(0)}{2} = N(0)e^{kt_{1/2}},$$

which directly follows from Eq. (2.1). Solving for k (do it) we find that the decay constant is given by

$$\boxed{k = -\frac{\ln 2}{t_{1/2}}},$$

where $\ln$ is in this instance the *natural logarithm* since we are dealing with the exponential e. Therefore, in the case of rubidium-87 (^{87}Rb)

$$k = \frac{-\ln 2}{49.15 \text{ billion (yr)}} = \frac{-0.6931}{0.4915 \times 10^{11} \text{(yr)}} = -1.41 \times 10^{-11}/\text{yr}.$$

We will use this value in the sequel.

Exercise

(a) Determine the decay constant for iodine-131. *Ans.* -0.0865/day.
(b) Determine the decay constant k for uranium-238 which has a half-life of 4.468×10^9 years. *Ans.* -1.551×10^{-10}/yr.

No rocks found on Earth have remained in their pristine state and have been subject to geologic processes since the Earth was formed. Fortunately, in order to determine the age of the Solar System we can take a sample from a meteorite that has come from the Asteroid Belt between Mars and Jupiter and *has* remained virtually unaltered since the time it was formed in the Solar Nebula.[3] Our meteorite sample will contain ^{87}Rb and also its decay product strontium-87 (^{87}Sr).

So let us now go back to Eq. (2.1) which can be applied to ^{87}Rb[4]:

---

[1] See Appendix 2 for a graph of the radioactive decay of carbon-11.
[2] This value has fluctuated slightly over the decades and is difficult to determine accurately. The above value is the one we have settled on.
[3] Not all meteorites were formed at the same time, but some are from the earliest period when the first solid material began to condense out of the Solar Nebula.
[4] Since in this section we are only dealing with radioactive decay, we could just as easily have taken the formula $^{87}Rb = {^{87}Rb_0}e^{-kt}$ in which case our constant k would be positive. À chacun son gout.

$$^{87}Rb = {^{87}Rb_0} e^{kt}, \qquad (2.2)$$

where $^{87}Rb_0$ represents the *initial amount* of ^{87}Rb when the sample formed. We do know that the constant $k = -1.41 \times 10^{11} (\text{yr})^{-1}$ determined above and we wish to solve for t since we also know the present amount of ^{87}Rb that can be measured in a lab. But what we do not know and is impossible to know is the initial amount of rubidium-87 ($^{87}Rb_0$) and so we need some other means to proceed further. But this is the only missing ingredient (besides t) so are not doing too badly at this stage.

The present amount of the decay product in our sample, ^{87}Sr, comes from both the initial amount that was there at the time of formation, which we denote by $^{87}Sr_0$, plus the amount from the long-term decay of ^{87}Rb, with the latter being represented by the quantity: $^{87}Rb_0 - {^{87}Rb}$. Mathematically this is just

$$^{87}Sr = {^{87}Sr_0} + \left(^{87}Rb_0 - {^{87}Rb}\right). \qquad (2.3)$$

It appears that we just made matters worse for ourselves since we now have another unknowable quantity to deal with, namely $^{87}Sr_0$. However, all is not lost as we now have two formulas (namely, Eqs. 2.2 and 2.3) both involving the unknown quantity $^{87}Rb_0$, so let us eliminate it by substituting its value from Eq. (2.2) into Eq. (2.3) to give

$$\begin{aligned} ^{87}Sr &= {^{87}Sr_0} + \left(^{87}Rb\, e^{-kt} - {^{87}Rb}\right) \\ &= {^{87}Sr_0} + {^{87}Rb}\left(e^{-kt} - 1\right). \end{aligned} \qquad (2.4)$$

Of course, this still leaves an unknowable quantity, the initial amount of strontium-87 ($^{87}Sr_0$).

At this juncture, we do something very clever and divide both sides of Eq. (2.4) by the *non-radioactive stable isotope* strontium-86 (^{86}Sr) which will also be found in our meteorite sample and has retained a constant value since the day the meteorite sample was formed. This gives us from Eq. (2.4),

$$\begin{aligned} \frac{^{87}Sr}{^{86}Sr} &= \frac{^{87}Sr_0}{^{86}Sr} + \frac{^{87}Rb\left(e^{-kt} - 1\right)}{^{86}Sr} \\ &= \left(e^{-kt} - 1\right)\frac{^{87}Rb}{^{86}Sr} + \frac{^{87}Sr_0}{^{86}Sr}, \end{aligned} \qquad (2.5)$$

after factoring the expression. Now the latter Eq. (2.5) is in the form of a straight line $y = ax + b$, where,

$$y = \frac{^{87}Sr}{^{86}Sr}; \quad a = \left(e^{-kt} - 1\right); \quad x = \frac{^{87}Rb}{^{86}Sr}; \quad \text{and } b = \frac{^{87}Sr_0}{^{86}Sr}.$$

The advantage of this straight-line formulation and the reason why we have taken this particular approach is that both of the ratios

$$x = \frac{^{87}Rb}{^{86}Sr}, \text{ and } y = \frac{^{87}Sr}{^{86}Sr},$$

can be measured via a mass spectrometer in the laboratory. Then we take several sample values and plot them in (x, y) coordinates and find the regression line for the data. This will give us the slope value $a = \frac{\Delta y}{\Delta x} = \left(e^{-kt} - 1\right)$ and from that we can solve for t, that is,

$$\boxed{t = -\frac{1}{k} \ln(a + 1)}, \tag{2.6}$$

which will be the age of the Solar System. Note that we avoided any consideration of the unknown quantity $^{87}Sr_0$ although the ratio $\frac{^{87}Sr_0}{^{86}Sr}$ is just the value at the y-intercept b.

Taking some real data from the Guareña H6 chondrite (Fig. 2.2), from Wasserburg et al., Initial strontium for a chondrite and the determination of a metamorphism or formation interval, *Earth and Planetary Sci. Lett.*, 7 (1) (1969), 33–43,

$$a = \left(e^{-kt} - 1\right) = 0.0665,$$

and taking our Eq. (2.6) for the time t above with $k = -1.41 \times 10^{-11} (\text{yr})^{-1}$,

$$t = \left(-\frac{1}{k}\right) \ln(a + 1) = \left(0.709 \times 10^{11}\right) \text{yr} \times \ln(1.0665) = 4.56 \times 10^9 \text{ years}.$$

This is the age of the Solar System including the age of the Sun.

It is hoped that the reader appreciates the thinking that goes into this approach as much as the author does, as we have achieved a rather remarkable result!

Moreover, the latest research using similar radiometric techniques on zircon crystals of a sample from Apollo 17 indicate that the age of the Moon is 4.46 billion years.[5] This is ~100 million years after the formation of the Earth and was most likely due to the impact of the Earth with a Mars-size impactor (known as Theia) which spewed debris from both bodies into a ring of material that coalesced into the Moon as we know it today.

---

[5] J. Greer et al., 4.46 Ga zircons anchor chronology of lunar magma, *Geochem. Persp. Lett.*, 27 (2023), 49–53.

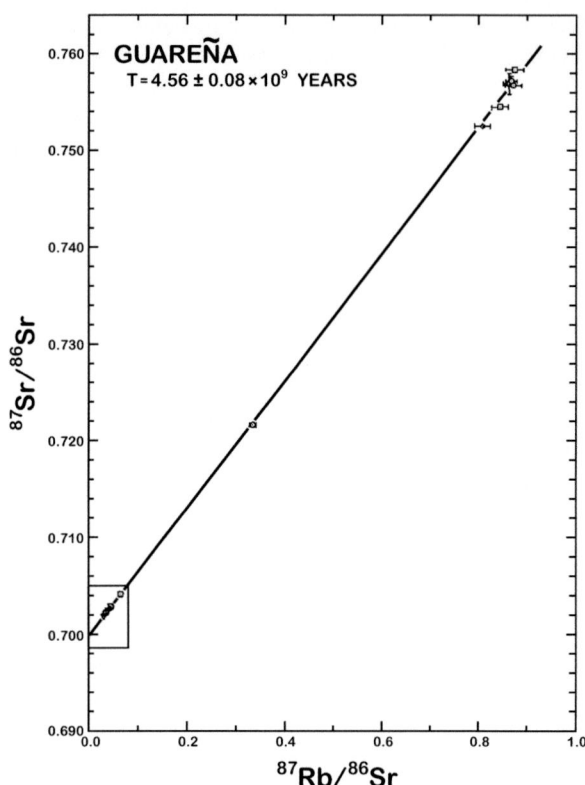

Fig. 2.2 The regression line (called an *isochron* since all values give the same time) from samples taken from the Guareña meteorite plotting the (x, y) coordinates of the ratios $x = \frac{^{87}Rb}{^{86}Sr}$ and $y = \frac{^{87}Sr}{^{86}Sr}$. The slope can be measured and comes out to $a = 0.0665$. (Courtesy G.J. Wasserburg et al., *Earth & Planet. Sci. Lett.* 7 (1969), 33–43/Elsevier)

Orbiting Earth

The formula for the orbital velocity of a body such as a satellite in a circular orbit of a much larger body such as the Earth is derived in Chap. 5 (Eq. (5.9)) and is given by

$$v_o = \sqrt{\frac{GM}{R}},$$

where G is the gravitational constant $G = (6.67430 \times 10^{-11})$ m³/kg·s², M is the mass of the large body being orbited and R is the *radius of the orbit* measured from the center of the massive body. Thus if h is the orbital altitude above the Earth, $R = R_e + h$, in the case of Earth orbiting satellites.

So let us consider the International Space Station (ISS) which is orbiting Earth at an average altitude of 420 km. The radius of the Earth is 6371 km so that the ISS is of a radius $R = 6371$ km + 420 km = 6791 km from the Earth's center. The mass of the Earth is $M_E = 5.972 \times 10^{24}$ kg. Therefore, from Eq. (5.9) noted above, we have

$$v_{ISS} = \sqrt{\frac{(6.674 \times 10^{-11} \text{m}^3/\text{kg} \cdot \text{s}^2)(5.972 \times 10^{24} \text{ kg})}{6.791 \times 10^6 \text{ m}}}$$

$$= 7.66 \times 10^3 \text{m/s},$$

or 7.66 km/s, which is much faster than a speeding bullet.

Escaping Earth

For any massive body such as the Earth, there is a minimum speed at which a projectile must travel in order to escape the gravitational pull of the body allowing it to travel indefinitely away from the body's surface. The equation for a body of mass M and of radius R will be derived in Chap. 4 (Eq. 4.14) and is (in the absence of atmospheric friction)

$$v_e = \sqrt{\frac{2GM}{R}}.$$

In the case of the Earth, whose radius is $R_E = 6.371 \times 10^3$ km, and whose mass is $M_E = 5.972 \times 10^{24}$ kg, we obtain

$$v_e = \sqrt{\frac{2 \times (6.674 \times 10^{-11} \text{ m}^3/\text{kg} \cdot \text{s}^2) \times 5.972 \times 10^{24} \text{ kg}}{6.371 \times 10^6 \text{ m}}}$$

$$= \sqrt{125.12 \times 10^6 \text{ m}^2/\text{s}^2}$$

11.19 km/s.

Observe that the formulas for orbital velocity and escape velocity look very similar but in the former case the value of R is the orbital radius from the center of the Earth whereas in the latter case R is simply the radius of the Earth. And of course, there is the factor 2 in the latter expression.

Exercise Determine the escape velocity of the Moon. Its mass and radius are given in Chap. 1. *Ans.* 2.38 km/s.

Mass of Earth

How does one weigh the Earth? There are several methods but perhaps the simplest stems from Newton's Law of Universal Gravitation discussed in Chap. 4. It is shown there that a direct consequence is the equation for the acceleration due to the Earth's gravity of a freely falling object (*sans* air resistance), namely (Eq. 4.2)

$$\boxed{g = \frac{GM_E}{R_E^2}}. \qquad (2.7)$$

where G is the gravitational constant, M_E and R_E are the mass and radius of the Earth respectively. The value of g can be measured by a device called an *accelerometer* at various points on the Earth's surface and an average taken. A round number figure is $g = 9.81$ m/s^2.

Knowing *a priori* the radius of the Earth R_E which can be deduced from its circumference $C_E = 2\pi R_E$, where $R_E = 6371$ km (again an averaged value), then we can work out the mass of the Earth,

$$M_E = \frac{gR_E^2}{G}$$

$$= \frac{(9.81 \text{ m/s}^2)(6.37 \times 10^6 \text{m})^2}{6.67 \times 10^{-11} \text{m}^3/\text{kg} \cdot \text{s}^2}$$

$$= 5.97 \times 10^{24} \text{ kg},$$

which is the (rounded-off) IAU accepted value.

One can also determine the mass of the Earth via orbiting satellites by solving for the mass in the equation appearing in the above discussion of orbital velocity,

$$v_o = \sqrt{\frac{GM}{R}}$$

so that

$$M = \frac{v_o^2 R}{G}.$$

Exercise Determine the mass of the Earth from the orbital velocity of a geostationary satellite[6] having velocity 3.07 km/s at an altitude of 35,786 km above the Earth's surface. *Ans.* 5.96×10^{24} kg.

---

[6]Geostationary satellites orbit the Earth once every 24 h and thus appear stationary above the Earth's surface which is useful for communication and meteorological purposes. They will be discussed again in Chap. 5.

Chapter 3
Let There Be Light

Electromagnetic Radiation

We perceive the Universe primarily through electromagnetic radiation which exhibits dual characteristics: it behaves both as a particle (a *photon*) and as a wave. Let us first consider the wave nature of light which is the same for all electromagnetic radiation such as radio/TV waves, visual light, X-rays, cellphone transmissions etc. Although sound waves require a medium for their transmission, such as air or water, electromagnetic waves can propagate through the vacuum of space. A main feature of these waves is their *wavelength,* which is the distance between any two peaks of the wave (see Fig. 3.1) or equivalently, between any two identically placed points on the wave itself. The wavelength symbol is usually denoted by the Greek letter lambda, λ, which represents the Greek 'el' sound.

All of these electromagnetic waves can be distinguished by their specific wavelengths which vary over a very wide spectrum. At very small wavelengths we have some common terminology based in terms of meters:[1]

meter—symbol m;
centimeter—**hundredths of a meter, 10^{-2} m, symbol cm;**
millimeter—**thousandths of a meter, 10^{-3} m, symbol mm;**
micrometer—**millionths of a meter, 10^{-6} m, symbol μm;**
nanometer—**billionths of a meter, 10^{-9} m, symbol nm.**

A common term for a micrometer is *micron* and is denoted by the Greek letter μ (mu) and it is also equal to one-thousandth of a millimeter. Therefore, 1 μ = 1 μm, and to give you some idea of its size, a typical human hair has a width between 17 and 180 μm.

---

[1] In Physics, there are even smaller quantities: *picometer* 10^{-12} m; *femtometer* 10^{-15} m. The diameter of an electron is about 10^{-14} m.

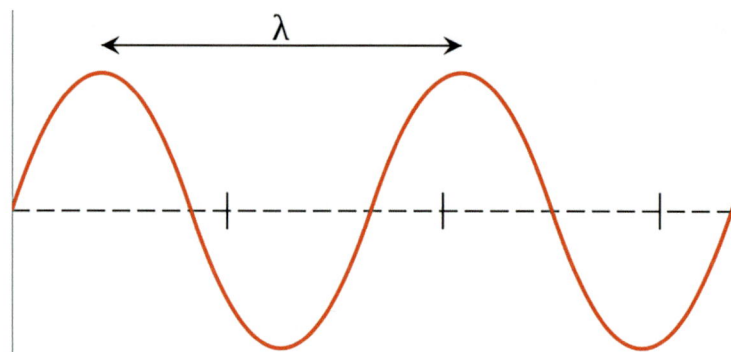

Fig. 3.1 The length of a wave depicted as the distance from one crest to another denoted by the symbol λ. (Public domain)

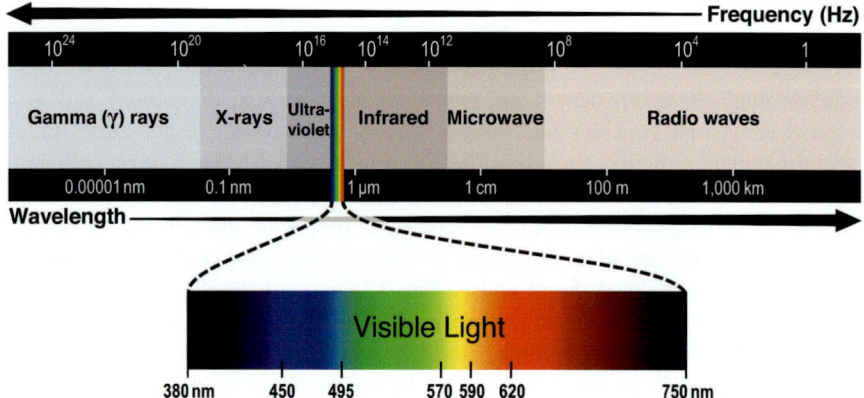

Fig. 3.2 The electromagnetic spectrum from gamma rays at the shortest wavelengths (highest frequencies) to radio waves in the longest wavelengths (lowest frequencies). Our eyes can only detect wavelengths in a very narrow band of visible light. (Courtesy Jonathan Park)

The wavelengths our eyes most interact with are those of the visible light spectrum and these vary from the violet end at 380 to 750 nm at the red end. Just below the violet range of the spectrum is the ultra[2] violet with wavelengths in the 10–380 nm range. Just above the red end of the spectrum is the infra[3] red with wavelengths ranging from 750 nm to 1 mm.[4] See Fig. 3.2.

However, wavelengths are just part of the story. All electromagnetic waves travel at the same speed, which is $c =$ **299,792,458 m/s** in a vacuum, called the *speed of light*. It is denoted by the letter c as it is a constant value throughout the Universe.

---

[2] In Latin, *ultra* means below.

[3] In Latin, *infra* is beyond.

[4] Various sources give slightly different wavelength values, but this is of no significance.

Moreover, it is constant even if the source of the waves is itself moving at great speed. This is somewhat counterintuitive since if you are on a train which is traveling at 100 km/h and you are walking on the train in the direction of its motion at 5 km/h, then your speed relative to the ground would be 105 km/h. If you walk in the opposite direction, then your speed relative to the ground is 95 km/h. But this is not the case with the speed of light.

If you are on a train traveling at, say 100,000 km/h and then shine a light from a flashlight either in the direction of the train's motion or in the opposite direction, the speed relative to the ground will remain c and unaffected by the motion of the train. This unusual feature has many interesting consequences, which appear in the discussion of the Theory of Relativity in Chap. 8.

For the sake of simplicity, we will often round off the speed of light to $c = 300{,}000$ km/s $= 3 \times 10^8$ m/s. Since the Earth is 40,075 km in circumference, this means that light could travel around the Earth $7\frac{1}{2}$ times in 1 s. As the Sun is roughly 150 million km on average from Earth, if we divide this by the speed of light we obtain 500 s, or $8\frac{1}{3}$ min for the light from the Sun to reach us. Thus, when a solar flare is observed on the Sun, it actually took place over 8 min prior to our observation.

Frequency

Another important aspect of electromagnetic waves is the notion of *frequency*, *f*, which is simply the number of complete waves (i.e., from crest to crest) that pass a given stationary point per second.[5] Each complete wave is considered a *cycle* so that the frequency measures how frequently the cycles are passing the reference point per second. Thus, the units of frequency are *cycles per second* denoted by the letters Hz, so that 60 cycles per second is written as 60 Hz.[6]

Often, we are dealing with a high number of cycles per second and denote:

10^3 cycles/s $=1$ kHz (1 kilohertz);
10^6 cycles/s $=1$ MHz (1 megahertz);
10^9 cycles/s 1 GHz (1 gigahertz);
10^{12} cycles/s 1 THz (1 terahertz).

As the frequency tells you how many cycles pass a given point in one second and the wavelength tells you the length of each individual cycle, then the product of the two quantities (frequency × wavelength) gives the total length the wave travels in one second. This distance per second is just the velocity of the wave which we already know to be the speed of light, c. Mathematically,

---

[5] The Greek symbol ν (nu) is also commonly used to indicate frequency.
[6] The letters Hz are used to honor the German physicist Heinrich Hertz, the first person to transmit radio waves.

$$c = f \times \lambda.$$

For example, if we consider red light having a wavelength $\lambda = 700$ nm $= 700 \times 10^{-9}$ m, then we can compute its frequency as: $f = 3 \times 10^8$ m/s $\div 700 \times 10^{-9}$ m $= 4.3 \times 10^{14}$ Hz $= 430$ THz. Note we are rounding off to keep matters simple.

Energy

Electromagnetic waves have energy that is proportional to their frequency/wavelength. The equation for this is given by

$$E = hf = h\frac{c}{\lambda}, \tag{3.1}$$

where $h = 6.626\,070\,15 \times 10^{-34}$ m$^2 \cdot$kg/s, is *Planck's constant*.[7] The units: m$^2 \cdot$ kg/s, are often replaced by J $\cdot$ s, where the unit of energy is in *Joules* (J).

This constant was introduced by Max Planck in 1900[8] in order to explain blackbody radiation (see below) which ultimately led to the development of the quantum theory. The above equation is significant for the fact that it shows that the energy of any form of electromagnetic radiation can only take on discrete multiples h of the light's frequency f. Heretofore, it was believed that electromagnetic energy was continuous, however Planck showed that this was not the case and that it could only be emitted in packets of a particular size called *quanta*.

For example, as the wavelength of red light $\lambda = 7 \times 10^{-7}$ m, each photon has energy

$$E = hf = h\frac{c}{\lambda}$$

$$= \left(6.626 \times 10^{-34} \text{ m}^2 \cdot \text{kg/s}\right) \times \left(\frac{2.998 \times 10^8 \text{m/s}}{7 \times 10^{-7}\text{m}}\right)$$

$$= 2.84 \times 10^{-19} \text{ m}^2 \cdot \text{kg/s}^2$$

$$= 2.84 \times 10^{-19} \text{ J}.$$

---

[7] There are also occasions to use the *reduced Planck constant*: $\hbar = \frac{h}{2\pi} = 1.054\,571\,817$

[8] Zur Theorie des Gesetzes der Energieverteilung im Normalspectrum (On the Theory of the Energy Distribution Law of the Normal Spectrum), *Verhandl. Dtsc. Phys. Ges.* 2 (1900), p. 237.

Blackbody Radiation

A blackbody is a theoretical construct in Physics that is an object that absorbs all incident electromagnetic radiation (with no reflection) irrespective of frequency/wavelength and emits radiation in a continuous spectrum with the intensity of radiation at each wavelength determined solely by its temperature. The Sun exhibits a reasonably close spectrum to a true blackbody but with slight deviations caused by absorption and emission features of the solar atmosphere. The *cosmic microwave background* (CMB) radiation emitted some 380,000 years after the Big Bang also closely matches that of a blackbody at 2.7 K.

Regarding the visible part of the spectrum one can see from Fig. 3.3 that at high temperatures the light emitted is predominately in the blue range of the spectrum and at lower temperatures the light is more in the red/yellow range which accords with everyday experience.

Wien's Displacement Law

Observe from the graphs that the larger the peak temperature the smaller the wavelength and this reciprocal relationship is given by the empirically derived *Wien's Displacement Law*,

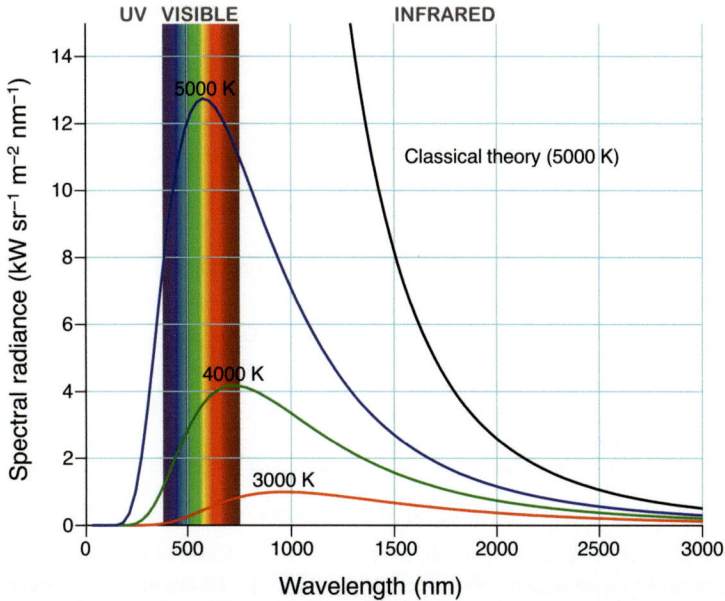

Fig. 3.3 Graphs of blackbody radiation for various temperatures. The black curve at right is the graph of the Rayleigh-Jeans formula discussed in the sequel. (Public Domain/Jonathan Park)

$$\boxed{\lambda_{max} = \frac{b}{T}} \qquad (3.2)$$

where $b = 0.0028977$ (m · K) is *Wien's constant* and T is the temperature measured in Kelvin (see Appendix 3). For example, taking a hot blue star with $T = 7000$ K,

$$\lambda_{max} = \frac{b}{T} = \frac{0.0028977 \text{ m} \cdot \text{K}}{7000 \text{ K}} = 414 \text{ nm}.$$

Obviously, rearranging Eq. (3.2), it is possible to determine the surface temperature of a star. This allows for stars to be placed into various *spectral classes* since their surface temperatures correlate with their color although other spectral features, such as certain absorption lines are also taken into account in a star's classification.[9]

Exercise Determine the surface temperature of the Sun if its peak wavelength is at 502 nm. (This is in the green range! But we perceive the Sun to be yellowish-white due to a combination of factors). *Ans.* 5772 K.

Planck's Law

For a more complete mathematical description of the intensity of radiation emitted by a blackbody at different wavelengths λ and at different temperatures T, we have Planck's Law for the *spectral radiance*[10]

---

[9]The seven spectral classes of stars with their approximate temperatures:

O-type: >30,000 K. Color blue. These are very hot stars.
B-type: 10,000 – 30,000 K. Color blue-white. They are still very hot.
A-type: 7500 – 10,000 K. Color white.
F-type: 6000 – 7500 K. Color yellow-white.
G-type: 5000 – 6000 K. Color yellow, like our Sun.
K-type: 3500 – 5000 K. Color orange.
M-type: <3500 K. These stars are red and the most prevalent (76.5%) of main sequence stars.

Note that different sources will have slightly different ranges for each class.

[10]The term *radiance* is measured in SI units of $W \cdot sr^{-1} \cdot m^{-2} = W/(sr \cdot m^2)$ and describes the power emitted/reflected by a surface per unit solid angle per unit projected area in a specific direction. For *spectral radiance* which further specifies the wavelength, the SI units are $W/(sr \cdot m^2 \cdot nm)$. Although the 'specific direction' is not explicitly specified it implies there is one, typically perpendicular to the surface of the emitter such as a star.

Blackbody Radiation

$$B(\lambda, T) = \frac{2hc^2}{\lambda^5} \cdot \frac{1}{e^{\frac{hc}{\lambda k_B T}} - 1} \quad (3.3)$$

where λ is the wavelength in nm, c the velocity of light in m/s, h is Planck's constant, k_B the *Boltzmann constant* $= 1.380\ 649 \times 10^{-23}$ J/K[11] and K is the temperature in Kelvin.[12]

The units of spectral radiance are $W \cdot sr^{-1} \cdot m^{-2} \cdot nm^{-1}$ and *represents the amount of energy emitted (or reflected) per unit steradian per unit area per unit of wavelength*. It represents how much energy is emitted per unit wavelength from each square meter of a blackbody surface (such as a star), in a specific direction per unit solid angle.

Since for a fixed temperature T, Planck's Law becomes a function of wavelength only, it should be possible to differentiate $B(\lambda, T)$ as a function of λ in order to find the value of λ_{max}. Indeed it is, and the details are found in Appendix 3 as they are not as straightforward as it might appear.

Wien's Approximation Law

There were two prior versions of Planck's Law, the first was given by Wien and prior to the introduction of Planck's constant. The one given by Wien can be rephrased however in terms of the Planck constant. It is a consequence of the fact that the exponential part of Planck's Law can be approximated by (see Appendix 4)

$$\frac{1}{e^{\frac{hc}{\lambda k_B T}} - 1} \approx e^{-\frac{hc}{\lambda k_B T}}$$

which yields *Wien's Approximation Law*

$$B(\lambda, T) = \frac{2hc^2}{\lambda^5} e^{-\frac{hc}{\lambda k_B T}}. \quad (3.4)$$

---

[11] The Boltzmann constant gives the relationship between the average kinetic energy of a gas particle and the temperature of the gas.

[12] Planck's Law can also be expressed in terms of frequency f and temperature T,

$$B(f, T) = \frac{2hf^3}{c^2} \cdot \frac{1}{e^{\frac{hf}{k_B T}} - 1}.$$

This formulation follows from the fact that $\lambda = \frac{c}{f}$, $|B(\lambda, T)d\lambda| = |B(f, T)df|$ and $d\lambda = -c/f^2\, df$, so that $B(f, T) = B(\lambda, T)\left|\frac{d\lambda}{df}\right| = B(\lambda, T)\frac{c}{f^2} = \frac{2hc^2}{c^5/f^5} e^{-\frac{hc}{(c/f)k_B T}}\left(\frac{c}{f^2}\right) = \frac{2hf^3}{c^2} \cdot \frac{1}{e^{\frac{hf}{k_B T}} - 1}$.

Rayleigh-Jeans Law

Another explanation of blackbody radiation that preceded that of Max Planck can also be derived from Planck's Law quite simply by considering the Taylor series

$$e^x = 1 + x + \frac{x^2}{2!} + \frac{x^3}{3!} + \dots$$

implying that

$$e^x \approx 1 + x$$

by dropping off all the higher powers of x for *small* values of $x \ll 1$. Then, as per the form in Eq. (3.3),

$$\frac{1}{e^x - 1} \approx \frac{1}{x},$$

again, for small x. From Eq. (3.3), then for *large* λ,[13] the quantity $x = \frac{hc}{\lambda k_B T}$ will indeed be small and thus we have from the above approximation

$$B(\lambda, T) = \frac{2hc^2}{\lambda^5} \cdot \frac{1}{e^{\frac{hc}{\lambda k_B T}} - 1}$$

$$\approx \frac{2hc^2}{\lambda^5} \cdot \frac{1}{hc/\lambda k_B T},$$

and simplifying we obtain[14]

$$\boxed{B(\lambda, T) = \frac{2ck_B}{\lambda^4} T.} \tag{3.5}$$

Voilà, we have just derived the *Raleigh-Jeans Law* which approximates Planck's Law for large values of the wavelength λ.

A comparison between the Wien Approximation Law, Planck's Law, and the Raleigh-Jeans Law is depicted in Fig. 3.4 for various values of a fixed temperature T.

---

[13] Equivalently, low frequency f.

[14] In terms of frequency, f, the Rayleigh-Jeans formula is: $B(\lambda, T) = \frac{2f^2 k_B}{c^2} T$.

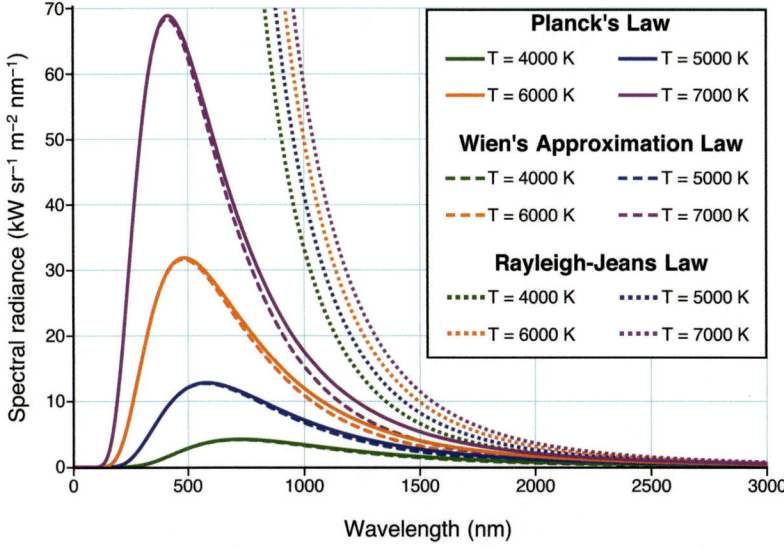

Fig. 3.4 A comparison between Planck's Law (solid lines), the Wien Approximation Law and the Raleigh-Jeans Law. Wien is very good for small values of the wavelength λ and Raleigh-Jeans comes into its own for large values of λ. (Courtesy Jonathan Park)

Ultraviolet Catastrophe

No one can discuss the Raleigh-Jeans Law without mentioning the very exotically named *Ultraviolet Catastrophe*. This immediately becomes apparent when examining the Raleigh-Jeans graph for a fixed temperature T as in Fig. 3.4 (and the Classical Theory curve in Fig. 3.3). As the graph approaches the shorter wavelengths (higher frequencies) of ultraviolet radiation in the 300 – 400 nm range, the values of the spectral radiance $B(\lambda, T)$ are becoming absurdly large. This is evident from the Eq. (3.5) as there is only a λ term in the denominator and so as λ gets arbitrarily small (equivalently, the frequency becomes arbitrarily large), the value of $B(\lambda, T)$ will become arbitrarily large. This obviously does not happen with a blackbody and so the resultant error in the Rayleigh-Jeans formulation was deemed the ultraviolet catastrophe. Such a colorful name for such a glowing error.

The 'catastrophe' does not occur with the Planck formulation since regarding the two terms of the Eq. (3.3)

$$\frac{2hc^2}{\lambda^5} \cdot \frac{1}{e^{\frac{hc}{\lambda k_B T}} - 1}$$

the first term does become infinite as λ becomes arbitrarily small. However, the second term is shrinking to zero at the same time and actually shrinking to zero at a *faster rate* than the first term is becoming infinite, with the result that the overall behavior of the product is to tend to zero with diminishing wavelength as in Fig. 3.4.[15]

Stefan-Boltzmann Law

Another blackbody radiation law describes the total energy radiated by a blackbody across *all wavelengths* (or frequencies) per unit of time per unit of surface area (W/m^2) and is known as *emittance*. It can also be derived from Planck's Law and says that the total energy is proportional to the fourth power of the temperature (in K), that is,

$$\boxed{E = \sigma T^4}, \tag{3.6}$$

where $\sigma = \frac{2\pi^5 k_B^4}{15 c^2 h^3} = 5.670\,374 \times 10^{-8}\,\text{W} \cdot \text{m}^{-2} \cdot \text{K}^{-4}$ is the *Stefan-Boltzmann constant*, and k_B, c, and h, are the usual suspects from above. Equation (3.6) is known as the *Stefan-Boltzmann Law*.[16] In Appendix 5 we show how it can be deduced from Planck's Law although historically it was derived independently of Planck's Law. The units are W/m^2 and if we multiply this by the surface area of a star, $S = 4\pi R^2$, where R is the radius of the star (in m), then we obtain the *total energy output per unit of time* of the star across all wavelengths

$$\boxed{L = 4\pi R^2 \sigma T^4} \tag{3.7}$$

where L is the *luminosity* in terms of watts W of power (J/s). For a star, the temperature T represents the *effective temperature*, that is, the temperature of a blackbody that corresponds to the same energy or luminosity given off by the star as per Eqs. (3.6) and (3.7).

The reader should appreciate the grandiosity of the constant σ which is composed of a lot of special ingredients. We have the ubiquitous π which comes from circles, the Boltzmann constant k_b from gas dynamics, the velocity of light c, and h from Quantum Mechanics. Adjoining the fourth power K^4 makes the fourth power of T in Kelvin cancel in the units.

For example, let us compute the total power output of the Sun which has radius $R = 6.9634 \times 10^8$ m, $\sigma = 5.670\,374 \times 10^{-8}\,\text{W} \cdot \text{m}^{-2} \cdot \text{K}^{-4}$, and the temperature is 5.773×10^3 K. Then by Eq. (3.7)

---

[15]This is a consequence of the fact that $\left(\frac{1}{x^n}\right)\left(\frac{1}{e^{1/x}-1}\right) \to 0$ as $x \to 0$ by L'Hôpital's Rule from Calculus. (Substituting $t = 1/x$ and letting $t \to \infty$ simplifies the calculation).

[16]The law holds for ideal blackbodies but for ones that are less than ideal, an additional factor called the *emissivity*, ϵ, is used, where ϵ varies between 0 and 1, with a perfect blackbody having $\epsilon = 1$.

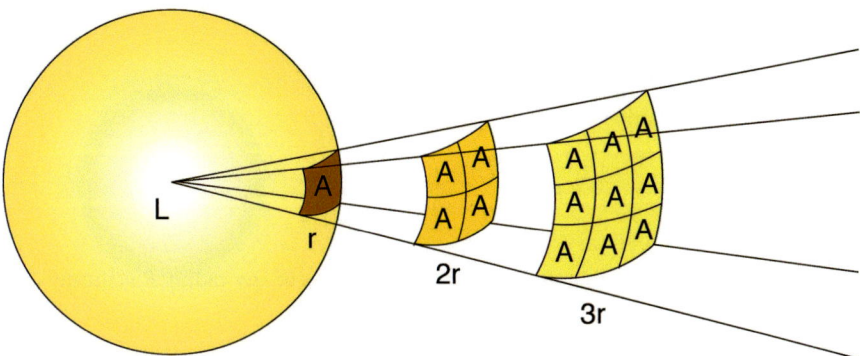

Fig. 3.5 The 3-dimensional geometry results in the intensity per unit area of a point source diminishing by a factor of $1/r^2$ at an arbitrary radius r from the source. A unit of area is indicated by the curvilinear patch A on the sphere of radius r, but at a distance of $2r$ the unit of area is now $4A$, and at a radius of $3r$, the unit of area becomes $9A$. (Courtesy Katy Metcalf)

$$L = 4\pi (6.9634 \times 10^8 \text{ m})^2 \times (5.670\,374 \times 10^{-8} \text{W} \cdot \text{m}^{-2} \cdot \text{K}^{-4}) \times (5.772 \times 10^3 \text{ K})^4$$

$$= 3.835 \times 10^{26} \text{ W}.$$

That can power a lot of 100 W light bulbs although a lot of the Sun's energy is lost into space and not received on Earth.

Inverse Square Principle

Several forces in Nature obey an inverse square law, such as the electrostatic force between two electric charges, the intensity of radiation such as light and sound, the magnetic force, and of course gravity. To see that this is a natural consequence of Geometry, suppose that a point source of radiation has luminosity L as in Fig. 3.5. On a sphere of arbitrary radius r from a point source having luminosity L, since the area of the sphere is given by

$$S = 4\pi r^2,$$

then the luminosity intensity per unit area or *radiant flux* F is (in W/m^2) measured at a distance r is

$$F = \frac{L}{4\pi r^2}. \tag{3.8}$$

Then, in view of Fig. 3.5, at a distance of $2r$, the flux is given by

$$F = \frac{L}{4\pi r^2} \cdot \frac{1}{4} = \frac{L}{4\pi(2r)^2},$$

and the flux at a distance of $3r$ is

$$F = \frac{L}{4\pi r^2} \cdot \frac{1}{9} = \frac{L}{4\pi(3r)^2},$$

and so forth. This shows that the intensity per unit area at any radius R is diminished by a factor of $1/R^2$

$$\boxed{F = \frac{L}{4\pi R^2}}, \tag{3.9}$$

which is the *inverse square principle*. This means that if we double the distance, the intensity is one-quarter that of the original distance which explains the rapid diminution of intensity with distance. The same holds for the brightness of a star in that a star at double the distance from an observer will appear to be one-quarter as bright as the nearer star.

A gravitational force works in the same way as will be explored in Chap. 4 as do other forces such as the electrostatic force and magnetic force.

Irradiance/Radiant Flux Density

The concept of *irradiance* (or *radiant flux density*) in Astronomy is a measure of the luminous power of a star incident on a unit area of a detector and in view of Eq. (3.9) can be expressed by[17]

$$\boxed{F = \frac{L}{4\pi D^2}} \tag{3.10}$$

where D is the distance to the star with the units in W/m^2. In other words, the radiant flux falls off as the square of the distance to the star as one expects.

The term *spectral irradiance* is the power per unit area *per unit wavelength* received by an observer. This has units W/m^2/nm and typically is obtained by dividing the irradiance F over a specific wavelength interval.

---

[17] The symbol b is also used for (apparent) brightness with $b = \frac{L}{4\pi D^2}$ and interchangeable with irradiance, radiant flux density, and flux, all measured in W/m^2.

Example For our Sun, $L = 3.837 \times 10^{26}$ W and the average distance is $D = 149.598 \times 10^9$ m, so that its radiant flux density is given by[18]

$$F = \frac{3.837 \times 10^{26} \text{ W}}{4\pi \left(149.598 \times 10^9 \text{ m}\right)^2}$$

$$\sim 1364 \text{ W/m}^2.$$

This value for F is known as the *Solar Constant*, although various sources report slightly differing values for it, some higher, some lower, than our value.

Exercise Compute the irradiance of the star Sirius which is at a distance of 2.64 parsecs and has a luminosity 25 times that of our Sun, more specifically, $L = 25.4 L_\odot$. Ans. $\sim 1.17 \times 10^{-7}$ W/m^2.

Hertzsprung-Russell Diagram

Commonly known as the *H-R diagram*, it is a visual depiction of the Stefan-Boltzmann Law in terms of luminosity as given by Eq. (3.7) above, whereby the luminosity of a star is a function of its temperature and size, as applied to various classes of stars. It was conceived around 1910 independently by the two astronomers Ejnar Hertzsprung (whom we will encounter later in Chap. 6 regarding Cepheid variables) and Henry Norris Russell and is central to the concept of stellar evolution. The temperature is derived from the Wien Displacement Law (Eq. (3.2)) and the luminosity can be derived from its proxy, the absolute magnitude via the relation[19]

$$\boxed{M = -2.5 \log\left(\frac{L}{L_\odot}\right) + M_\odot}, \quad (3.11)$$

where $L_\odot$ and $M_\odot$ are the luminosity and absolute magnitude of the Sun respectively given in Chap. 1. To see this, note that if the distance to two stars is exactly the same, say d_0, then as per footnote 17,

$$\frac{b_1}{b_2} = \frac{L_1}{L_2}.$$

---

[18] As we are taking the average distance to the Sun, this flux value will vary throughout the year.

[19] Essentially the same relation is used in dealing with galaxies with $L_\odot$ and $M_\odot$ replaced by the corresponding data of a standard reference galaxy as is done in Chap. 6 regarding the Tully-Fisher relation.

where b_1, b_2 and L_1, L_2 are the brightness and luminosity of the two stars respectively. And when considering the *absolute magnitudes* of two stars, we do indeed have the same distance, $d_0 = 10$ pc. Moreover, the absolute magnitudes of the two stars M_1, M_2 obey the same magnitude relationship as in Eq. (1.7) (since $m = M$ at $d_0 = 10$ pc) implying that[20]

$$\frac{L_1}{L_2} = 100^{\left(\frac{M_2 - M_1}{5}\right)}.$$

Taking the logarithm of both sides gives us

$$\log \frac{L_1}{L_2} = \frac{2}{5}(M_2 - M_1),$$

and rearranging the terms with the Sun as the second comparison star yields the relation given by Eq. (3.11).

For example, taking a supergiant with absolute magnitude $M = -5.17$, and $M_\odot = 4.83$, Eq. (3.11) gives

$$-5.17 = -2.5 \log \left(\frac{L}{L_\odot}\right) + 4.83,$$

so that $\log\left(\frac{L}{L_\odot}\right) = 4$ and consequently $L = 10{,}000\, L_\odot$ in accordance with Fig. 3.6.

From the H-R diagram we can note the following features:

Comments on H-R Diagram

(i) Stars begin life on the main sequence and remain there while their core is fusing hydrogen into helium. The position along the main sequence is primarily determined by the star's mass. The most massive, hot and luminous ones are on the upper left of the main sequence and the least massive, cool and less luminous are at the lower right.

(ii) High mass stars evolve and leave the main sequence becoming giants and supergiants which pass through a series of nuclear burning stages becoming cooler and more luminous. In view of the Stefan-Boltzmann relation $L = 4\pi R^2 \sigma T^4$, if T decreases and L increases then the radius R must increase and this becomes very large for such cool stars.

(iii) Low to intermediate mass stars (like our Sun) eventually evolve into white dwarfs, which are the remnants of stars that have exhausted their nuclear fuel

---

[20] We are reversing the order of the indices 1 and 2 here from Eq. (1.7) for no reason in particular save to keep the reader alert.

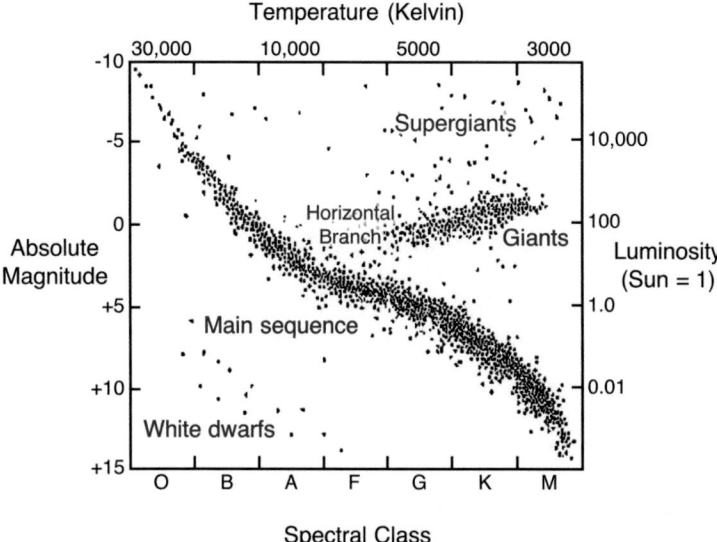

Fig. 3.6 Hertzsprung-Russell Diagram showing the four main categories of stars. Our Sun is of spectral class G to the right of center on the main sequence. (Courtesy NASA/CXC/SAO)

and collapsed into a small dense state. The latter are on the hot end of the scale but with low mass and luminosity. Again, $L = 4\pi R^2 \sigma T^4$ implies that in this scenario that a low luminosity and high temperature implies that the size of the star must be very small. Indeed, of comparable size to the Earth (see Fig. 8.9).

(iv) If two stars have the same luminosity but one has a higher temperature, then the latter must have a smaller radius in view of the Stefan-Boltzmann Eq. (3.7).

Example Suppose that a star the size of the Sun has a luminosity L that is 100 times that of the Sun. Then to find the star's temperature T observe that

$$\frac{L}{L_\odot} = 100 = \left(\frac{T}{T_\odot}\right)^4,$$

implying that $T = T_\odot \sqrt{10} = 18{,}253$ K, taking the value $T_\odot = 5772$ K.

Exercise Find the temperature of a star that is the size of the Sun and has a luminosity L that is 1000 times that of the Sun. *Ans.* 32,458 K.

Eddington Luminosity

One question that is reasonable to ask is just how luminous can a star be? Can we just increase the values of R and T in the Stefan-Boltzmann relation? The answer is no, as something has to give at some critical stage known as the *Eddington Luminosity* or *Eddington Limit*.

In order for a star to maintain its structural stability it has to perform a balancing act between the inward tug of gravity and the outward pressure of its radiation. This relationship between gravity and radiation pressure was investigated by Sir Arthur Eddington early in the twentieth century. The formula that now bears his name describes the limiting luminosity of a star that can still be in equilibrium, namely

$$\boxed{L_{Edd} = \frac{4\pi G M c m_p}{\sigma_T}}, \qquad (3.12)$$

where M is the mass of the star and the only new parameters here are m_p, the mass of a proton and σ_T is the so-called *Thomson scattering cross-section* for an electron. The mass of the proton features here as protons constitute the main mass of the star[21] and σ_T gives the cross-sectional area over which a photon of light can scatter off an electron. This cross-section is naturally very small: $\sigma_T = 6.652 \times 10^{-29} \mathrm{m}^2$ and the mass of the proton is: $m_P = 1.673 \times 10^{-27}$ kg.

Let's try it for the Sun. Then

$$L_{Edd} = \frac{4\pi \times (6.674 \times 10^{-11} \mathrm{\ m^3/kg \cdot s^2}) \times (1.989 \times 10^{30} \mathrm{kg}) \times (2.998 \times 10^8 \mathrm{m/s}) \times (1.673 \times 10^{-27} \mathrm{\ kg})}{6.652 \times 10^{-29} \mathrm{m}^2}$$

$$= 1.26 \times 10^{31} \mathrm{m}^2 \cdot \mathrm{kg/s}^3$$

$$= 1.26 \times 10^{31} \mathrm{\ W}.$$

As we have now done the messy calculation for the Sun, we can now by taking the ratio of a general L_{Edd} to our L_{Edd} value for the Sun, express Eq. (3.12) in the much simplified form (since the constants in both cancel out)

$$\boxed{L_{Edd} = 1.26 \times 10^{31} \left(\frac{M}{M_\odot}\right) \mathrm{W}.} \qquad (3.13)$$

If we compare this with the actual luminosity of the Sun which is $L_\odot = 3.837 \times 10^{26}$ W, we find that the Eddington luminosity for the Sun is roughly 32,800 times the Sun's current luminosity. But this all begs the question: What if a

---

[21] The mass of the electron is not very significant compared to that of a proton and is ignored and we are ignoring helium and other possible star constituents and only considering the simplest case of the star being predominantly hydrogen.

star had a smaller companion star that it was continually drawing mass off of? What happens when the star reaches the Eddington luminosity? It should be mentioned however, that the value of the Eddington luminosity limit will be affected by other factors if the star is not primarily hydrogen.

There are several possibilities. The radiation pressure could blow away some of the accreting material through powerful *stellar winds* in the outer layers that would prevent the star from exceeding the Eddington luminosity. Or the accretion of new material could become highly variable through losing some mass and then the accretion of further mass. Another interesting scenario is the formation of an *accretion disk* surrounding a white dwarf, neutron star, or black hole, parts of which can become so hot and luminous and generating sufficient radiation pressure as to also act as a deterrent to the accretion of further material. This can indirectly lead to the formation of jets or outflows.

Sérsic Profile[22]

Another quantity related to radiance is *surface brightness intensity*, which is the amount of light per unit area from an object such as a galaxy or nebula and represents the total amount of light per unit area on the plane of the sky. The units are usually in magnitude/arcsec^2 and can be measured at various wavelengths.

It is clear from images of galaxies that the surface brightness diminishes with increasing distance from the center. An empirical measure of surface brightness of galaxies conforms reasonably well with the formula of Sérsic represented by an exponential function

$$\boxed{I(R) = I_0 e^{-bR^{1/n}}} \tag{3.14}$$

where R represents the *apparent distance* from the center projected onto the plane of the sky[23] and $I_0 = I(0)$ is the surface brightness at the center of the galaxy, with n called the *Sérsic index*. For spiral galaxies we have $n = 1$ and

$$I(R) = I_0 e^{-bR},$$

whereas for elliptical galaxies we have $n = 4$ which was the original formulation of de Vaucouleurs (1948)

---

[22] Also called *Sérsic's Model*, or *Sérsic's Law* due to José Luis Sérsic, Influencia de la dispersión atmosférica e instrumental sobre las distribuciones de luminosidad en una galaxia, *Bol. Asoc. Argentina de Astronomía*, 6 (1963), 41–43. It is a generalization of a similar model given much earlier by Gerard de Vaucouleurs for elliptical galaxies, *Ann. Astrophys.* 11 (1948), 247.

[23] Note that this is not the same as the actual radial distance.

$$I(R) = I_0 e^{-bR^{1/4}}.$$

Actually, a more elaborate formulation is used by astronomers, that has been normalized by the notion of the *effective radius* (*half-light radius*), R_e that encloses one-half of the total light of the galaxy, and I_e the intensity of the light at the effective radius. This gives the formulation of the Sérsic profile as

$$\boxed{I(R) = I_e e^{\left(-b_n \left[\left(\frac{R}{R_e}\right)^{\frac{1}{n}} - 1\right]\right)}}, \tag{3.15}$$

or in logarithmic terms

$$\boxed{\log \frac{I}{I_e} = -\kappa_n \left[\left(\frac{R}{R_e}\right)^{\frac{1}{n}} - 1\right]} \tag{3.16}$$

where $\kappa_n = \frac{1}{\ln 10} b_n$.[24] Observe that the sign in brackets of Eq. 3.16 goes from positive to negative as the radius goes from $R < R_e$ to $R > R_e$ for any value of $n > 0$ and where b_n is a constant that depends on n. Clearly, at $R = R_e$, $I(R_e) = I_e$ as desired. In particular, for a spiral disk galaxy with $n = 1$, we have $b_1 \approx 1.678$ and

$$\log \frac{I}{I_e} = -\kappa_n \left[\left(\frac{R}{R_e}\right) - 1\right]$$

whereas for an elliptical galaxy with $n = 4$, $b_4 \approx 7.669$. For each value of n the constant κ_n (that is, b_n) is defined by the condition that $I(R_e)$ contains one-half the total light of the galaxy but the details will not be pursued here as they are somewhat arcane. However, a common approximation formula is given by ($n > 0.36$)

$$b_n \approx 2n - \frac{1}{3} + \frac{4}{405n} + \frac{46}{25{,}515n^2} + \frac{131}{114{,}8175n^3},$$

or more simply, $b_n \approx 2n - \frac{1}{3}$.

Exercise Determine the value of b_8. Ans. ≈ 15.668.

We need to say a word about the effective radius R_e. Unless the galaxy has circular symmetry when projected against the plane of the sky, then the contour line along which $I \equiv I_e$ will not be a true circle but rather a distorted one depending on the actual shape of the galaxy. Contour lines along which the light intensity remains constant are called *isophotes* so there is one specific isophote that corresponds to the effective radius (Fig. 3.7).

---

[24] This is a consequence of the fact that: $\log e^x = \left(\frac{1}{\ln 10}\right) x$, where *ln* is the natural logarithm base e.

Sérsic Profile

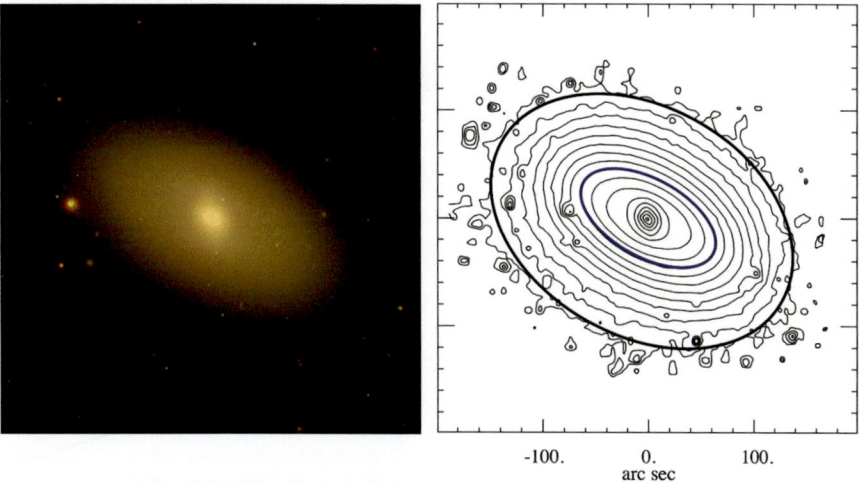

Fig. 3.7 (L): The lenticular galaxy NGC 3412; (R) the isophotes of constant light intensity. The drawn ellipses represent approximations for comparison purposes of the data. In this instance observe the change in eccentricity from the outer to inner regions. (Courtesy (left) Sloan Digital Sky Survey; (right) Peter Erwin, Max Planck Institute for Extraterrestrial Physics)

From the Fig. 3.8 we can glean several important features.

$n < 1$: Indicative of less concentrated disk galaxies with a gradual decline in intensity which falls off more rapidly for $R > R_e$.[25]

$n = 1$: Typical of spiral galaxies, the intensity decreases in an exponential manner with the radius.

$1 < n < 4$: Intensity decreasing faster than exponential decline with radius but not as fast as the de Vaucouleurs profile with $n = 4$.

$n \geq 4$: Typical of elliptical galaxies with a rapid decline in intensity beyond the core region and a higher intensity near the core with increasing values of n.

It should be pointed out that whenever the brightness of a celestial object is measured, the background sky brightness must be subtracted in order to obtain the true brightness of the object. Moreover, we mention that there are various other surface brightness profiles found in the astronomical literature as the light distribution from a galaxy can be quite complex but the Sérsic profile serves as a good exemplar.

---

[25] In particular, for $n = 0.5$,

$$I(R) = I_e e^{\left(-0.695\left[\left(\frac{R}{R_e}\right)^2 - 1\right]\right)}.$$

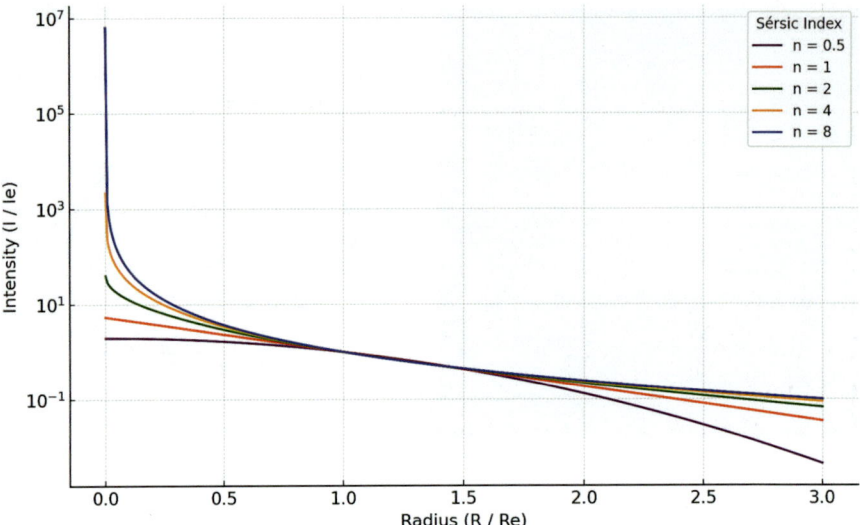

Fig. 3.8 The Sérsic profile for differing values of the index n. The x-axis represents the radius R normalized by the effective radius R_e and the *logarithmic* y-axis gives the intensity I normalized by the intensity at the effective radius, I_e, so that $I = I_e$ when $R = R_e$. Note that as n increases, the profile becomes much steeper approaching the galaxy's center, indicating a denser core that is typical for elliptical galaxies. Lower values of n show a more gradual decrease in intensity moving away from the core, characteristic of disk galaxies. Generated by GPT 4 by the author

Transit Method

One very interesting development over the last few decades is the discovery of planetary bodies (*exoplanets*) orbiting other stars in the Milky Way, mainly within a few thousand light-years due to the technical difficulties involved. During the orbit the exoplanet will transit its host star, and if the alignment with Earth is suitable, in so doing it will block some of the star's light. This decrease in light can be used to determine the existence of an exoplanet and measure its size. The *fractional decrease in light from the host star* is known as the *transit depth* denoted by δ and is related to the ratio of the areas of the two bodies.

Of course, the decrease in light as a result of an exoplanet must be distinguished from that of a variable star or eclipsing binary pair but there are various distinguishing features such as the shape of the dip in the light curve, its duration, and so forth. For example, exoplanets typically have a distinct symmetrical dip in the light curve with regular short period dips whereas variable stars can have longer more irregular patterns in their light curve. Eclipsing binaries can also be carefully teased out from exoplanets by the depth and shape of the dip in the light curve as a result of the size of the eclipsing star among other factors (Fig. 3.9).

Since area is proportional to the square of the radius of a circular body (which we assume to be the case), we have the formula for the transit depth

Transit Method

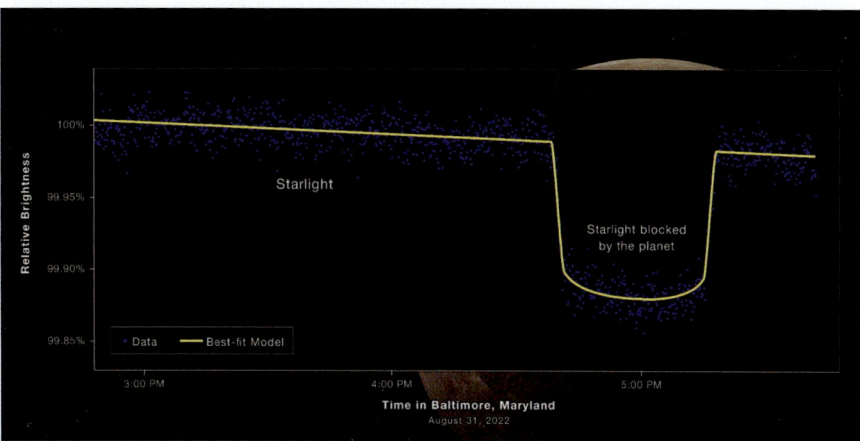

Fig. 3.9 The actual light curve of the exoplanet LHS 475b taken by the James Webb Space Telescope exhibiting the typical exoplanet diminution in the host star's light curve. The transit lasted about 40 min and the data includes 1158 individual brightness measurements (purple dots). (Courtesy NASA, ESA, CSA, Leah Hustak (STScI))

$$\delta = \left(\frac{R_p}{R_s}\right)^2,$$

where R_p is the radius of the exoplanet and R_s is the radius of the star. Solving for the exoplanet radius gives

$$R_p = R_s \sqrt{\delta}. \tag{3.17}$$

As an example, let us consider one of the first exoplanets discovered (1999), HD 209458b. The host star is HD 209458 which has a radius $R_s = 1.15 \times R_\odot = 1.15 \times 696{,}340$ km $= 800{,}791$ km. The transit depth has been measured at $\delta = 1.5\% = 0.015$. Therefore,

$$R_p = 800{,}791 \times \sqrt{0.015} = 98{,}076 \text{ km}.$$

For comparison, the radius of Jupiter is 69,911 km and so HD 209458b is a Jupiter-size exoplanet.

Exercise Determine the radius of the exoplanet TrES-2b that orbits a host star the same size as the Sun and yields a transit depth of 1.7%. *Ans.* 90,792 km.

This method only works when the exoplanet and host star are suitably aligned in the sky so that the transit can be viewed from Earth. Moreover, stars are not uniformly bright across their disk and exhibit a phenomenon called *limb darkening* whereby the outer stellar regions appear darker than at the center due to the decreased density and cooler layers of the star matter at the limb. Thus, if a star crosses near the edge of the star the transit depth will be reduced and this factor needs to be taken into account. Fortunately, there are various other methods as well as the transit method, including gravitational microlensing (see Chap. 8), and perturbations of the star's position for detecting exoplanets among others.

Chapter 4
Newton's Laws

Among the many wonderous scientific discoveries and pronouncements in Isaac Newton's *Philosophiæ Naturalis Principia Mathematica,* published in 1687, were his laws of motion which initiated the modern rigorous study of Physics and Astronomy.

Newton's Three Laws of Motion

1. (*Law of Inertia*) *A body remains in a state of rest or in uniform motion in a straight line unless acted upon by an external force.*

The Law of Inertia has been further elaborated upon by the first postulate of the Special Theory of Relativity in Chap. 8.

2. *The rate of change of a body's momentum is equal to the force acting upon it.*

The momentum of a body of mass m moving with velocity v, is defined to be

$$p = mv,$$

or more properly, in vector notation, $\bar{p} = m\bar{v}$, since velocity and momentum are vectors, that is, they have both a magnitude and direction. Knowing this fact, we will generally use the non-vector notation for simplicity. The letter p is historical and we just have to go with the flow. If the mass m is in kilograms (kg) and the velocity v is in meters per second (m/s), then the momentum is expressed in the units kg · m/s.

Then the rate of change of momentum (with respect to time) is dp/dt and the Second Law says that if F is the force acting on the body, then

$$F = \frac{dp}{dt} = m\frac{dv}{dt}$$
$$= ma,$$

since the rate of change of the velocity, dv/dt, is the acceleration a. So, the Second Law boils down to $\boldsymbol{F = ma}$.

3. *Whenever one body exerts a force on another body, the second body exerts an equal force on the first body in the opposite direction to the first force. In other words, for every action there is an equal and opposite reaction.*

Note that the action and reaction are applied to two different bodies. For example, when a tennis player bounces a ball on the ground (the action), the ground forces the ball (the reaction) to bounce back up. A book lying on a table is attracted by the Earth and the book in turn is attracting the Earth with an equal and opposite force.

As a consequence of the Third Law we have the principle of the *Conservation of Momentum*:

For two (or more) bodies acting upon each other in a closed system, the total momentum of all the bodies will remain constant.

To see this, we need only consider two bodies acting upon each other with forces F_1 and F_2 respectively. By Newton's Third Law, $F_1 = -F_2$ which implies that regarding their respective momenta, p_1, p_2,

$$0 = F_1 + F_2 = m_1 a_1 + m_2 a_2$$
$$= \frac{dp_1}{dt} + \frac{dp_2}{dt}$$
$$\frac{d}{dt}(p_1 + p_2).$$

From elementary Calculus, if the derivative of a function equals zero, then the function is constant and thus,

$$p_1 + p_2 = \text{constant}.$$

Gravity

According to the Newtonian view of the world, gravity is a force that is akin to the intensity of a light source at a point in that it diminishes as the square of the distance from the source. Newton formulated this notion in *Principia* as the *Law of Universal Gravitation*: *Every particle in the Universe attracts every other particle with a force that is proportional to the product of their masses and inversely proportional to the*

Gravity

distance between their centers. Restating this mathematically, if m_1 and m_2 are the two masses and r is the distance between their centers, then

$$F = \frac{Gm_1m_2}{r^2}, \qquad (4.1)$$

where G is the *gravitational constant* given in Chap. 1, namely $G = 6.67430 \times 10^{-11}$ m^3/kg $\cdot$ s^2. The constant G is one of the fundamental constants of Nature and is encountered often in the study of the Universe.

In the particular case of the Earth where $r = R_E = 6371$ km, then Eq. (4.1) becomes

$$F = \frac{GM_E m}{R_E^2} = m\left(\frac{GM_E}{R_E^2}\right) = mg, \qquad (4.2)$$

where $g = GM_E/R_E^2$ is the acceleration of a falling body in the Earth's gravitational field, neglecting air resistance. A value of 9.81 m/s^2 is usually sufficient and a more precisely measured value will depend on the exact location on the Earth's surface where the measurement is taken.[1] The value indicates that for every additional second of a body in free-fall above the Earth, it will gain an additional 9.81 m/s in velocity. From Eq. (4.2) we determined the mass of the Earth via Eq. (2.7).

Exercise Determine the acceleration g_m of a freely falling body on the Moon. The requisite mass and radius are in Chap. 1. *Ans.* 1.625 m/s^2.

It was also proved by Newton in the *Principia* that a spherically symmetric body gravitationally attracts every other body as if its entire mass were concentrated at its center.[2] For this reason, *we assume throughout that all bodies with mass are spherically symmetric*. This makes gravity related calculations that much simpler as one only has to take into account the distance between the centers of the two bodies and not their physical size.

In vector form recall that any nonzero vector v can be made into a *unit vector* $\bar{u} = \bar{v}/\|\bar{v}\|^3$ of length 1. If we have a mass m_1 at location given by the vector $\bar{r_1}$ and mass m_2 at location $\bar{r_2}$ then the vector $\bar{r_2} - \bar{r_1}$ is a vector from m_1 to m_2 in the direction of m_2. So, the unit vector from m_1 to m_2 in the direction of m_2 is given by

---

[1] The standardized value is $g = 9.80665$ m/s^2.
[2] This is Newton's *Shell Theorem*.
[3] The notation $\|v\|$ is used to indicate the length of the vector v which is the distance from the origin to the tip of the vector. Thus $\|r_2 - r_1\|$ represents the length of the vector $r_2 - r_1$ and the latter is in the direction of the vector r_2.

Fig. 4.1 Vector diagram indicating the force $\overline{F_{12}}$ exerted on the mass m_1 by the mass m_2. (Courtesy Katy Metcalf)

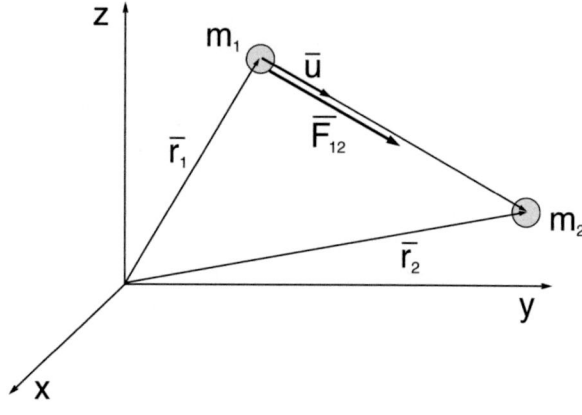

$$\overline{u} = \frac{\overline{r_2} - \overline{r_1}}{\|\overline{r_2} - \overline{r_1}\|}$$

and this will give the *direction* of the force exerted on mass m_1 by the mass m_2. Since it is a unit vector it does not perturb the magnitude of any scaler values it appears with (Fig. 4.1).

Then the force exerted on mass m_1 by the mass m_2 is given by

$$\overline{F_{12}} = \frac{G m_1 m_2}{\|\overline{r_2} - \overline{r_1}\|^2} \cdot \frac{\overline{r_2} - \overline{r_1}}{\|\overline{r_2} - \overline{r_1}\|}$$

$$= \frac{G m_1 m_2}{\|\overline{r_2} - \overline{r_1}\|^3} (\overline{r_2} - \overline{r_1}).$$

We can simplify this a little by setting $r = r_2 - r_1$ which gives

$$\overline{F_{12}} = \frac{G m_1 m_2}{\|\overline{r}\|^3} \overline{r},$$

where $\overline{r}$ is the vector from m_1 to m_2 in the direction of m_2.

Of course, there is the gravitational force $\overline{F_{21}}$ exerted on mass m_2 by the mass m_1, so that the vector $\overline{r}$ is in the opposite direction, that is, $\overline{r} = \overline{r_1} - \overline{r_2} = -(\overline{r_2} - \overline{r_1})$ and this implies that $\overline{F_{21}} = -\overline{F_{12}}$.

In principle, the radius can be arbitrarily large which means that the Earth exerts a gravitational attraction upon every body in the Universe and vice versa. However, in practice, the magnitude of the force between very distant bodies is infinitesimal and insignificant due to the inverse square diminution, and so can effectively be ignored.

The acceleration of a small body of mass m due to the gravity of a much larger body of mass M can now easily be put into vector form by taking a unit vector $\overline{r}$ directed outward from the center of the larger body (effectively the origin) to give

Gravitational Potential Energy

$$\bar{g} = -\frac{GM}{R^2}\bar{r}.$$

Exercise Determine the gravitational acceleration of a freely falling body on the planet Mercury. The mass and radius of Mercury are $M = 3.3011 \times 10^{23}$ kg, and $R = 2.4397 \times 10^6$ m, respectively. Ans. $g = 3.70$ m/s², a bit over one-third that on Earth.

Gravitational Potential Energy

Gravity can also be used to do work by storing it as potential energy in heavy masses. This *gravitational potential energy* of an object of mass m at a distance R from the Earth's center is given by

$$\boxed{U = -\frac{GMm}{R}} \qquad (4.3)$$

and represents the amount of work done by the gravitational attraction of a large mass M to bring a smaller mass m from infinity to a particular point with R being the distance between their centers.[4]

There can be a simplification in the case of the Earth's surface and a mass at an elevation h above it. Then the (positive) difference in gravitational potential is given by[5]

$$\Delta U = \frac{GMm}{R_E} - \frac{GMm}{R_E + h} = GMm\left(\frac{1}{R_E} - \frac{1}{R_E + h}\right)$$
$$= GMm\frac{h}{R_E(R_E + h)}.$$

Since R_E is the radius of the Earth, in our case, $h \ll R_E$ so that $R_E + h \approx R_E$, which implies that

$$\Delta U \approx GMm\frac{h}{R_E^2}.$$

---

[4] This work is given by $W = \int_\infty^R F dr = \int_\infty^R \frac{GMm}{r^2} dr = -\frac{GMm}{r}\Big|_\infty^R = -\left(\frac{GMm}{R} - \lim_{r\to\infty}\frac{GMm}{r}\right) = -\frac{GMm}{R}$.

[5] Note that here we are considering: $-\frac{GMm}{R_E+h} - \left(-\frac{GMm}{R_E}\right)$ which is positive.

The quantity $\frac{GM}{R_E^2} = g$ is a constant and taking $U = 0$ on the Earth's surface by convention, just as a reference point in this setting for convenience, we have a general expression for the gravitational potential of a mass m at height h,

$$\boxed{U = mgh}. \tag{4.4}$$

This formula is currently being exploited as a means to store energy (via a *gravity battery*) by lifting a large mass to a considerable height off the ground using not needed surplus energy (or solar power) and then gradually lowering the weight to convert the gravitational potential energy of the mass to kinetic energy in order to generate electricity when needed (Fig. 4.2).

We can also see via Eq. (4.3) that the *total energy* of a small body of mass m having orbital velocity v_o around a larger body of mass M will consist of its kinetic energy[6] plus its potential energy and is given by

$$E_{tot} = \frac{1}{2}mv_o^2 - \frac{GMm}{R} = \frac{1}{2}m\frac{GM}{R} - \frac{GMm}{R}$$

where we have invoked Eq. (5.8) for the orbital velocity v_o to obtain the last expression and therefore,

$$\boxed{E_{tot} = -\frac{1}{2}\frac{GMm}{R}}.$$

As a consequence, particles in orbit around a black hole with a tremendous mass M will have commensurate enormous energies.

We can also describe the gravitational energy of a uniformly dense sphere of radius R and a mass M which is given by (see Appendix 6)

$$\boxed{U = -\frac{3GM^2}{5R}}. \tag{4.5}$$

This is a very useful result and plays an important role in the sequel relating to the collapse of a large cloud of gas under gravitational contraction which can lead to the formation of the stars including our Sun.

---

[6]The kinetic energy of a body of mass m moving with velocity v is given by $K = \frac{1}{2}mv^2$.

Fig. 4.2 What goes up, must come down. A gravity battery constructed by Energy Vault with each concrete block weighing 35 tons but other more sustainable materials are planning to be used. (Courtesy Energy Vault)

Virial Theorem

The Virial Theorem is a basic principle of Physics governing the distribution of kinetic energy and potential energy in a stable system of particles. Here, particles can mean gas molecules, stars in a galaxy, galaxies in a cluster, and there is even a version in Quantum Mechanics. As a consequence, it has applications in a wide scientific arena such as astrophysics, thermodynamics, celestial mechanics, among others. Indeed, it was Fritz Zwicky in 1933 who used the Virial Theorem to determine the existence of dark matter discussed in Chap. 5.

The Virial Theorem says that the average kinetic energy of a system of a stable system is equal to minus one-half the average potential energy, that is,

$$\boxed{K = -\frac{1}{2}U}. \tag{4.6}$$

The theorem can be mathematically deduced but it has also been experimentally verified. It represents the requisite relationship between kinetic and potential energy for the system to remain stable.

Since the Virial Theorem can be applied to a cloud of gas, now let us ask: Under what conditions will a molecular cloud of gas begin to collapse under its own gravity to form a star? There are many clouds of gas and dust in the Universe that have not collapsed to form stars so there must be certain criteria that need to be met for star formation to occur. And such conditions were address by James Jeans in 1902 with the result of the James Criterion.

Jean's Criterion

To simplify matters let us assume that the molecular hydrogen gas (H_2) cloud[7] is a sphere of some uniform density ρ. We begin by writing the Virial Theorem of Eq. (4.6) in the form

$$2K + U = 0. \tag{4.7}$$

For the potential energy, we have the expression given by Eq. (4.5) for a molecular cloud of mass M and radius R,

$$U = -\frac{3GM^2}{5R}.$$

For the kinetic energy K, we need to borrow an equation from the kinetic theory of gases which describes the kinetic energy of the cloud in terms of its temperature and how many particles it contains,

---

[7] There will also be present other molecules and atoms in the cloud besides H_2, for example, helium, carbon monoxide, methane, ammonia, even water molecules, but we ignore their presence for our calculation

Virial Theorem

$$K = \frac{3}{2} N k_b T$$

where N is the number of molecules in the cloud, k_b is the Boltzmann constant, and T is the temperature of the cloud measured in kelvin.[8] The quantity N is awkward to work with so we can eliminate it since it equals the total mass of the cloud M divided by the mass of each molecule m,[9] that is

$$N = \frac{M}{m},$$

so that the kinetic energy can be written as

$$K = \frac{3M}{2m} k_b T.$$

Putting the two expressions for U and K into Eq. (4.7), gives us

$$\frac{3M}{m} k_b T = \frac{3GM^2}{5R} \quad (4.8)$$

where the kinetic energy of the cloud due molecular motion is represented by the left-hand side and the potential energy by the right-hand side. In this case we have an equilibrium state. If we actually want the molecular cloud to collapse, we require the gravitational potential energy to *exceed* the kinetic energy of outward pressure, that is in Eq. (4.8) we require[10]

$$\frac{3M}{m} k_b T < \frac{3GM^2}{5R}.$$

Cancelling a couple of terms leads to

---

[8] According to the *Equipartition of Energy Law*, the average kinetic energy per molecule in a monatomic (single atom per molecule) ideal gas is $\frac{3}{2} k_b T$ and our gas cloud contains N molecules. We use this formulation in spite of the fact that our gas is diatomic H_2 not monatomic without going into any further required technicalities. With this understanding our results are very good approximations of the physics involved. We further assume that there are no other factors responsible for the kinetic energy such as turbulence.

[9] Sometimes this is expressed is μm_H where m_H is the mass of the hydrogen atom and μ is some constant. In our case we are taking $\mu = 2$ and $m = 2m_H$ for simplification.

[10] If the kinetic energy exceeds the gravitational potential energy (say, due to its temperature T) and we have the reverse inequality then the gas cloud will expand.

$$\frac{k_b T}{m} < \frac{GM}{5R}. \tag{4.9}$$

Since the cloud is spherical, we have an expression for the mass in terms of its volume which is

$$M = \rho V = \rho \cdot \frac{4}{3}\pi R^3. \tag{4.10}$$

We simply have to insert this expression for the mass M into the preceding Eq. (4.9) to eliminate the mass and obtain

$$\frac{k_b T}{m} < \frac{4\pi \rho G R^3}{15R}. \tag{4.11 prelim}$$

Finally, solving for the radius R,

$$\boxed{R > \sqrt{\frac{15 k_b T}{4\pi \rho G m}} = \lambda_J}. \tag{4.11}$$

The square root quantity λ_J is called the *Jeans length* and the radius of the gas cloud should be greater than this value for the cloud to collapse under gravity.

But what about the mass of the cloud? It too must exceed a certain threshold. We simply go back to Eq. (4.10) and write the radius R in terms of mass M,

$$R = \left(\frac{3M}{4\pi \rho}\right)^{1/3}$$

and putting this expression into the Jeans length Eq. (4.11) for the value R,

$$\left(\frac{3M}{4\pi \rho}\right)^{1/3} > \left(\frac{15 k_b T}{4\pi \rho G m}\right)^{1/2}.$$

If we cube both sides and apply a bit of arithmetic[11] then

---

[11] After cubing both sides and solving for M note that $\frac{(15)^{3/2}}{3} = 5^{3/2} \cdot 3^{1/2}$ and $\frac{4\pi \rho}{(4\pi \rho)^{3/2}} = \frac{1}{(4\pi \rho)^{1/2}}$. Combining the square root terms gives the term $\left(\frac{3}{4\pi \rho}\right)^{1/2}$ and the $5^{3/2}$ combines with the other terms of the same power.

Virial Theorem

$$\boxed{M > \left(\frac{5k_bT}{Gm}\right)^{\frac{3}{2}}\left(\frac{3}{4\pi\rho}\right)^{\frac{1}{2}} = M_J.} \tag{4.12}$$

Here the term M_J is the *Jeans mass*.

Both the Jeans length and Jeans mass derivations represent a beautiful application of the Virial Theorem. There is another approach to deriving λ_J involving the speed of *sound crossing-time* of the gas cloud but that can be left for another day. Note that squaring both sides in Eq. (4.12) allows us to simply interchange the positions of ρ and M^2 so that if we know the mass of the star then its density must be

$$\boxed{\rho > \left(\frac{5k_bT}{Gm}\right)^{3}\left(\frac{3}{4\pi M^2}\right) = \rho_J,} \tag{4.13}$$

which is the *Jeans density*.

Example Let us use Eq. (4.12) to compute the Jeans mass of a diffuse molecular hydrogen gas cloud at a (typical) temperature of $T = 40$ K and a density of $\rho = 6.6 \times 10^{-19}$ kg/m^3. The value of the mass of the hydrogen molecule is $m = 2m_H = 3.32 \times 10^{-27}$ kg, so that

$$M_J = \left(\frac{5k_bT}{Gm}\right)^{3/2}\left(\frac{3}{4\pi\rho}\right)^{1/2}$$

$$= \left(\frac{5 \cdot 40 \text{ K} \cdot 1.38 \times 10^{-23} \text{m}^2 \cdot \text{kg/s}^2 \cdot \text{K}}{(6.67 \times 10^{-11} \text{m}^3/\text{kg} \cdot \text{s}^2)(3.32 \times 10^{-27} \text{kg})}\right)^{3/2}\left(\frac{3}{4\pi \cdot 6.6 \times 10^{-19} \text{kg/m}^3}\right)^{1/2}.$$

Considering just the units, after a bit of cancelling in each expression we will have

$$\left(\frac{\text{kg}}{\text{m}}\right)^{3/2}\left(\frac{\text{m}^{3/2}}{\text{kg}^{1/2}}\right) = \text{kg},$$

just as we expect. For the numerical part of the calculation, we obtain

$$M_J = (1.25 \times 10^{16})^{3/2}(36.2 \times 10^{16})^{1/2} = 8.4 \times 10^{32} \text{kg}.$$

This amounts to ~422 $M_\odot$.

Exercise Compute the number of molecules per cubic centimeter of the giant molecular cloud in the preceding Example. *Ans.* ~200.

Exercise
(a) Assume a giant molecular H_2 cloud has a temperature $T = 20$ K and a mass of 1000 $M_\odot$. Compute the Jeans density. *Ans.* 1.46×10^{-20} kg/m^3.

(b) How many H_2 molecules does this represent per cubic centimeter? *Ans.* ~4 H_2 molecules per cm^3.

Escape Velocity

In order for a body of mass m to escape the gravitational attraction on the surface of a body of mass M such as the Earth or other planet having radius R, the mass m must have a certain minimum velocity v_e. This velocity can be determined since the kinetic energy of the mass m must be at least equal and opposite in sign to that of the gravitational potential energy at the surface of the larger body, that is, in view of Eq. (4.3)

$$\frac{1}{2}mv_e^2 = \frac{GMm}{R},$$

and solving for v_e,

$$\boxed{v_e = \sqrt{\frac{2GM}{R}}}. \qquad (4.14)$$

This is the formula used in Chap. 2 to determine the velocity required to leave planet Earth.

Exercise Determine the escape velocity of Mars which has a mass of $M = 64.169 \times 10^{22}$ kg and a radius of $R = 3389.5$ km. *Ans.* 5.027 km/s.

The notion of orbital velocity such as a satellite orbiting a planet will be dealt with in the next chapter (Eq. (5.9)).

Roche Limit

The Roche limit describes how close a small satellite that is held together by its own self-gravity can approach a much larger 'primary' mass so that the tidal forces of the primary mass not result overcome the self-gravity of the smaller mass resulting in the latter being torn apart. Our presentation makes a number of simplifications yet still results in a very good approximation to the Roche limit.[12]

---

[12] These considerations were first taken up by French astronomer Édouard Roche in 1848 in a series of papers.

Fig. 4.3 Depiction of the tidal forces exerted on a satellite of mass m by a primary mass M. The tidal force is the differential force between the gravitational force F_1 on the near face minus the gravitational force F_2 on the far side. (Courtesy Katy Metcalf)

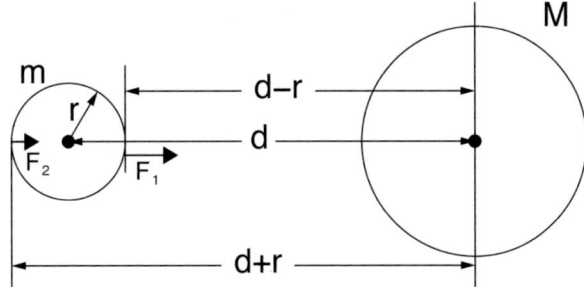

Tidal Forces

In the first instance, in order to dissect this situation, let us first consider the tidal forces on a satellite of mass m and radius r that is a distance d (between their centers) from a much larger mass M as in Fig. 4.3.

The *tidal force* acting on the satellite will be the difference between the gravitational pull on the near side minus the gravitational pull on the far side. For the near side the gravitational force acting on it is

$$F_1 = \frac{GMm}{(d-r)^2},$$

and for the far side we have

$$F_2 = \frac{GMm}{(d+r)^2}$$

Then the tidal force acting on the satellite is given by

$$F_T = F_1 - F_2$$
$$= \frac{GMm}{(d-r)^2} - \frac{GMm}{(d+r)^2}$$
$$= GMm \left[\frac{1}{(d-r)^2} - \frac{1}{(d+r)^2} \right].$$

We can simplify the term in the brackets by factoring out the d term to give

$$F_T = \frac{GMm}{d^2} \left[\frac{1}{(1-\frac{r}{d})^2} - \frac{1}{(1+\frac{r}{d})^2} \right]$$

which we write as

$$\boxed{F_T = \frac{GMm}{d^2}\left[\left(1-\frac{r}{d}\right)^{-2} - \left(1+\frac{r}{d}\right)^{-2}\right]}. \tag{4.15}$$

This now is the tidal force on the satellite due to the mass M. So far so good.

The terms in the brackets are a bit cumbersome to deal with so we can simplify them further via the algebraic binomial expansion:

$$(a+x)^n = a^n + na^{n-1}x^1 + \ldots + x^n,$$

where in our case $a = 1$, $n = -2$, with $x = -\frac{r}{d}$ and again with $x = \frac{r}{d}$. Thus the bracketed terms of Eq. (4.15) become

$$\left[\left(1 + 2\frac{r}{d} + \ldots\right) - \left(1 - 2\frac{r}{d} + \ldots\right)\right]$$
$$= 4\frac{r}{d} + \ldots.$$

The terms omitted by the ellipsis represent higher powers of $\frac{r}{d}$ and since the radius of the satellite will be very much smaller compared to d, making $\frac{r}{d}$ small, these higher powers can be omitted. Therefore, from Eq. (4.15) we arrive at

$$\boxed{F_T \approx \frac{4GMmr}{d^3}}. \tag{4.16}$$

This takes care of the tidal force. Our assumption is that the satellite is held together by self-gravity and to greatly simplify matters let us next consider the force binding the near-side hemisphere of the satellite with the far-side hemisphere. Furthermore, let us regard the two hemispheres to be in the form of two adjacent spheres each of mass $\frac{m}{2}$ and radius $\frac{r}{2}$ as in Fig. 4.4 left. The distance between the centers of the two spherical halves is still $r = \frac{r}{2} + \frac{r}{2}$, so that the gravitational force between the two spheres is

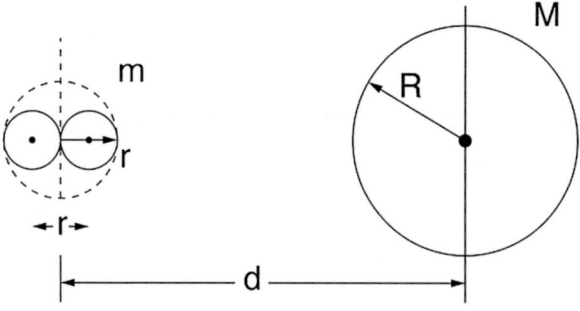

Fig. 4.4 The gravitational force binding the left side and right side of the satellite considered as two spheres, each of mass $m/2$ and radius $r/2$. (Courtesy Katy Metcalf)

Roche Limit

$$\boxed{F_G = \frac{G\left(\frac{m}{2}\right)\left(\frac{m}{2}\right)}{r^2} = \frac{Gm^2}{4r^2}}. \tag{4.17}$$

Our goal is to find the point where the tidal force F_T due to the distant mass M will exceed the gravitational force holding the two components of the satellite together F_G, that is, we require

$$F_G \leq F_T.$$

This is to say, in view of Eqs. (4.16) and 4.17,

$$\frac{Gm^2}{4r^2} \leq \frac{4GMmr}{d^3}.$$

Canceling a few terms yields,

$$d^3 \leq \frac{16Mr^3}{m},$$

and solving for d, the satellite must not approach any closer than the *Roche limit*,

$$\boxed{d \approx 2.5r\left(\frac{M}{m}\right)^{\frac{1}{3}}}. \tag{4.18}$$

This formula can also be written in terms of the density of the satellite and the distant primary mass as the density is the mass divided by its volume. In particular,

$$\rho_M = \frac{M}{\frac{4}{3}\pi R^3}, \qquad \rho_m = \frac{m}{\frac{4}{3}\pi r^3},$$

where R is the radius of the primary mass and solving each of the above equations for the respective masses,

$$M = \frac{4}{3}\pi R^3 \rho_M, \qquad m = \frac{4}{3}\pi r^3 \rho_m.$$

Putting both of these values into the Roche limit formula Eq. (4.18) gives

$$\boxed{d \approx 2.5R\left(\frac{\rho_M}{\rho_m}\right)^{\frac{1}{3}}}. \tag{4.19}$$

In spite of the unsophisticated nature of our analysis, various other estimates of much greater complexity also give a value close to the constant 2.5. Roche himself arrived at a value for the constant equal to 2.456. On the other hand, there are some moons of the outer planets found within the Roche limit due to other dynamical forces. For example, the moon Galatea orbiting Neptune lies within the Roche limit of Neptune's ring system. Or the small moon Pan, orbits within the Encke Gap of Saturn's A ring. So, one must tread lightly when discussing the Roche limit, but in general it is a useful guideline.

Example Let us take a 'moon' of Neptune composed of water ice so that:

$\rho_m = 0.917$ gm/cm^3;
$\rho_M = 1.64$ gm/cm^3, the average density of Neptune;
$R = 24{,}622$ km, the radius of Neptune.

Then the Roche limit for this 'moon' would be

$$d \approx 2.5R \left(\frac{\rho_M}{\rho_m}\right)^{\frac{1}{3}}$$

$$= 2.5 \times (24{,}622 \text{ km}) \left(\frac{1.64}{0.917}\right)^{1/3}$$

$$\approx 74{,}717 \text{ km}.$$

Note that this is the distance measured between the centers of both Neptune and our 'moon'.

Exercise In the case of the Neptunian moon Galatea, its density is: $\rho_m \approx 0.75 \frac{\text{gm}}{\text{cm}^3}$.

(a) Compute the Roche limit for Galatea. *Ans.* 79,896 km.
(b) The distance between Galatea and Neptune is 37,200 km (i.e., the distance between their surfaces). Galatea has a diameter of ≈ 87 km and the mean radius of Neptune is 24,622 km. Show that Galatea lies within its Roche limit.

Exercise The Saturnian moon Pan has a mean density of 0.4 gm/cm^3. Compute its Roche limit, assuming Saturn's mean radius is 58,232 km and its density is 0.687 gm/cm^3. *Ans.* 174,341 km. *Note:* Pan lies within the Encke division which is at a distance of 133,590 km from the center of Saturn so that Pan lies well within its Roche limit.

Chapter 5
Kepler's Laws

It was Copernicus who, in 1543 in his *De revolutionibus orbium coelestium* (*On the Revolutions of the Heavenly Spheres*) placed the Sun as the center of the Solar System with the planets revolving around it. Then in the early seventeenth century Johannes Kepler published his three laws governing the motion of the planets around the Sun. These were derived empirically from a wealth of astronomical data compiled by the Danish astronomer Tycho Brahe.

Ellipses

As the planets move around the Sun in elliptical orbits with the Sun, it behooves us to know some geometric properties about ellipses.[1]

An ellipse centered at the origin has *foci* at the points $(-c, 0)$ and $(c, 0)$, with the *semi-major axis* length given by a and the *semi-minor axis* length denoted by b. The *sum* of the lengths of the lines L_1 and L_2, drawn from each focus to any point on the ellipse, remains constant. This constant value, say k, can be determined when taking the point $(a, 0)$ on the ellipse so that

$$k = L_1 + L_2 = (a + c) + (a - c) = 2a,$$

so that $k^2 = 4a^2$.

Now, taking the point $(0, b)$, we see that $L_1 = L_2 = L$, with $2L = k$, so that $k^2 = 4L^2$. Also, from Fig. 5.1 and the Pythagorean Theorem,

---

[1] In the absence of any external forces such as other nearby planets, the elliptical orbits predicted by Kepler's first Law are closed in that the planetary body returns to the same point after each revolution. However, gravitational influences by external bodies as well as relativistic effects will cause the elliptical orbit to *precess*. See Chap. 8.

Fig. 5.1 The distances $L_1 + L_2 = $ constant in any ellipse. (Courtesy Katy Metcalf)

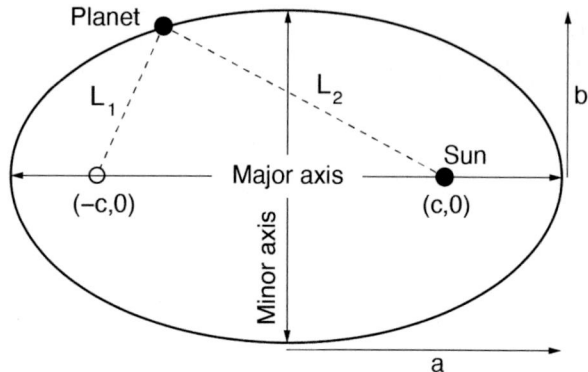

$$b^2 + c^2 = L^2.$$

Since $k^2 = 4a^2$ and $k^2 = 4L^2 = 4b^2 + 4c^2$, we obtain, $a^2 = b^2 + c^2$, that is, we can write the loci in terms of the semi-major and semi-minor axes as,

$$\boxed{c^2 = a^2 - b^2}. \tag{5.1}$$

A distinguishing characteristic is that of *eccentricity* which measures how much the ellipse deviates from a circle:

$$\boxed{e = \frac{c}{a}}, \tag{5.2}$$

so that the length of each focus is $c = ea$ as in Fig. 5.2.

Note that in the case of a circle, the two foci meet into one point at the origin so the $e = 0$. In the case of the Earth, the eccentricity has a value of $e = 0.01671$ which is very low and hence the Earth's orbit around the Sun is nearly circular, whereas Pluto's orbit is more elliptical with $e = 0.25$. From Eq. (5.1) we have

$$e = \frac{\sqrt{a^2 - b^2}}{a} = \sqrt{1 - \frac{b^2}{a^2}},$$

which defines the eccentricity strictly in terms of the semi-major and semi-minor axes.[2]

---

[2] The equation in the xy-plane of an ellipse centered at the origin is given by

$$\frac{x^2}{a^2} + \frac{y^2}{b^2} = 1,$$

where a and b are the respective semi-major and semi-minor axes respectively and (x, y) is any point on the ellipse. Again, when $a = b$ the ellipse becomes a circle which is a special case of an ellipse.

Ellipses

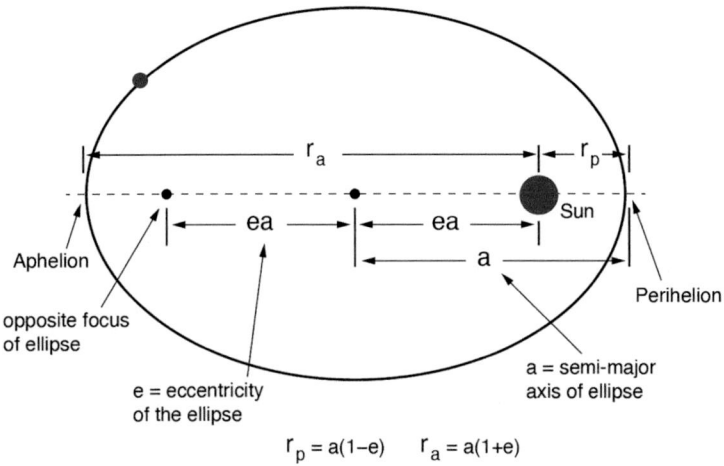

Fig. 5.2 Various components that make up the elliptical orbits of celestial bodies. (Courtesy Katy Metcalf)

For a planet in our Solar System, the *perihelion* is the closest point to the Sun in the planet's orbit and the *aphelion* is the furthest point from the Sun in the planet's orbit. More generally, the corresponding points for any small body orbiting a larger one are called *periapsis*, r_p, and *apoapsis*, r_a, respectively (as in Fig. 5.2). Then their distances are given by

$$r_a = a + ea = a(1 + e),$$
$$r_p = a - ea = a(1 - e).$$

If these two distances are known then we can obtain the semi-major axis very conveniently since we can add the two preceding equations to give a very useful formula

$$\boxed{a = \frac{r_a + r_p}{2}}. \tag{5.3}$$

As an example, periodic comet Encke has aphelion $r_a = 4.098$ AU and perihelion $r_p = 0.3396$ AU so that its semi-major axis is

$$a = \frac{4.098 \text{ AU} + 0.3396 \text{ AU}}{2} = 2.2188 \text{ AU}.$$

Furthermore, the eccentricity can be calculated from r_a and r_p, since dividing their two respective formulas,

$$\frac{r_a}{r_p} = \frac{1+e}{1-e},$$

and solving this equation for e gives us

$$e = \frac{r_a - r_p}{r_a + r_p}. \tag{5.4}$$

For example, the apoapsis of the Earth is $r_a = 152.10 \times 10^6$ km and the periapsis is $r_p = 147.10 \times 10^6$ km, so that the eccentricity is (*sans* units since they cancel)

$$e = \frac{152.10 \times 10^6 - 147.10 \times 10^6}{152.10 \times 10^6 + 147.10 \times 10^6} = 0.01671,$$

as was mentioned above.

The formula in polar coordinates for the ellipse itself can also be derived simply in terms of its eccentricity, and semi-major axis, *if we put the right focus at the origin*[3]

$$r(\theta) = \frac{a(1-e^2)}{1 + e \cos \theta}, \tag{5.5}$$

where $r(\theta)$ is the distance from the origin to the point (r, θ) on the ellipse.[4] The algebraic details are of no concern to us here but note that in the case of a circle and $e = 0$ we have $r(\theta) = a$ as it should be. Furthermore, when $\theta = 0$, $r(\theta) = a(1-e) = r_p$, and when $\theta = 180°$ then $r(\theta) = a(1+e) = r_a$.

---

[3] If one puts the left focus at the origin then the equation has a minus sign in the denominator.
[4] The formula in polar coordinates for $r(\theta)$ allows for the calculation of the *average distance* from a point on the ellipse to a focus via

$$\overline{r(\theta)} = \frac{1}{2\pi} \int_0^{2\pi} \frac{a(1-e^2)}{1 + e \cos \theta} d\theta = a\sqrt{1-e^2}.$$

This follows from the evaluation of the integral

$$\int_0^{2\pi} \frac{d\theta}{1 + e \cos \theta} = \frac{2\pi}{\sqrt{1-e^2}},$$

which is not particularly straight-forward and need not be pursued here. For orbits with small values of eccentricity, the average distance is approximately a.

Ellipses

Fig. 5.3 The galaxy M 59 (NGC 4621) classified as E5 in view of the classification Eq. (5.6a). (Courtesy Wikisky.org)

Ellipticity

A related concept to eccentricity is that of *ellipticity* which is the measure

$$\boxed{E = 1 - \frac{b}{a}} \quad (5.6)$$

where a and b are the semi-major and semi-minor axes respectively. Note that when $a = b$ we have a circle and $E = 0$. As an example we can take the galaxy M 59 (NGC 4621) in Fig. 5.3 which in a very crude fashion the author using a ruler measured the ratio $\frac{b}{a} = 0.5$ and therefore $E = 0.5$.

This value of the ellipticity can give a convenient quantitative classification for elliptical galaxies, namely

$$\boxed{EX = 10\left(1 - \frac{b}{a}\right)} \quad (5.6a)$$

where X represents the value of the numerical calculation on the right. Consequently, the elliptical galaxy M 59, is classified as an E5 galaxy. See Fig. 5.3. The *Hubble Classification* goes from E0 where $b = a$ and the galaxy is essentially circular to E7 where b is only $0.3a$ and the shape is very elongated. Of course, real elliptical galaxies are not so perfectly constrained.

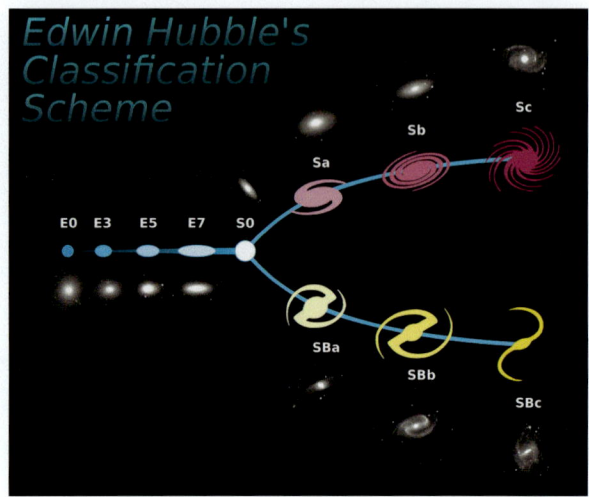

Fig. 5.4 The Hubble Classification Scheme also known as Hubble's Tuning Fork Diagram depicting the morphology of the major galaxy types. (Credit: ESA/NASA)

Expanding on the elliptical galaxy classification there is an elaborate classification scheme devised by Edwin Hubble in 1926 based on a galaxy's appearance often referred to as *Hubble's Tuning Fork Diagram*. Hubble included four classes: elliptical, spiral, barred spirals, and irregulars. The scheme is strictly based on morphology and does not imply any sort of evolution of one type into another although spirals can evolve into lenticular galaxies (SO) and elliptical galaxies can result from the merger of spiral galaxies which can violently disrupt their spiral structures. Irregular galaxies were placed in a separate category not included in the tuning fork scheme, and needless to say that the scheme has been revised and expanded many times over the years (Fig. 5.4).

Orbital Elements

When one body orbits another there are a number of distinct features that characterize the orbit due to the geometry of the orbit. As will be seen in the next section, orbiting bodies travel in elliptical paths and therefore there is a number of various determinants that go into a complete description of the orbit's parameters. We have already discussed the ellipse features of semi-major and semi-minor axes along with the eccentricity but there are other parameters used to characterize the elliptical orbit of a celestial body.

For a start we need a reference plane and taking our Solar System for the present illustration, we use the plane of the Earth's orbit, known as the *ecliptic plane* or *plane*

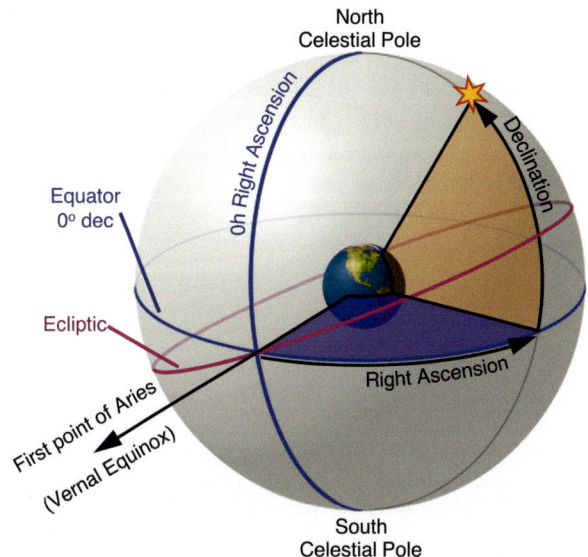

Fig. 5.5 The first point of Aries is a reference direction in space where the Sun has crossed the equatorial plane from south to north at the vernal equinox indicated by the arrow. The point now lies in the constellation of Pisces but historically was in the constellation of Aries and retains the name. (Courtesy Jonathan Park)

of the ecliptic.[5] While the other planets have orbits that lie close to the ecliptic, they orbit in planes slightly tilted to the ecliptic and this tilt is called the *inclination* (i).

Here are the approximate inclinations for the planets:

Mercury: 7.00°
Venus: 3.39°
Mars: 1.85°
Jupiter: 1.31°
Saturn: 2.49°
Uranus: 0.77°
Neptune: 1.77°.

Clearly the orbit of Mercury has the greatest inclination and Uranus has the smallest, the latter nearly orbiting in the ecliptic. One the other hand, dwarf planet Pluto has an inclination of about 17° to the ecliptic plane and another dwarf planet, Eris, which is roughly the same size as Pluto, is inclined at an even greater angle at a whopping 44°.

Furthermore, for determining the location in space of any celestial body there is a very standard *reference direction* that is determined by the line of intersection of the Earth's equatorial plane with the ecliptic plane at the time of the vernal equinox as in Fig. 5.5. This occurs in the northern hemisphere on ~20 March when the apparent path of the Sun crosses the celestial equator from south to north and results in equal day and equal night. The position of the Sun in the sky where this occurs is called the *first point of Aries* denoted by the symbol ♈ and is the location of 0° right ascension

---

[5] The term *ecliptic* is the apparent path of the Sun as viewed from Earth over the course of a year.

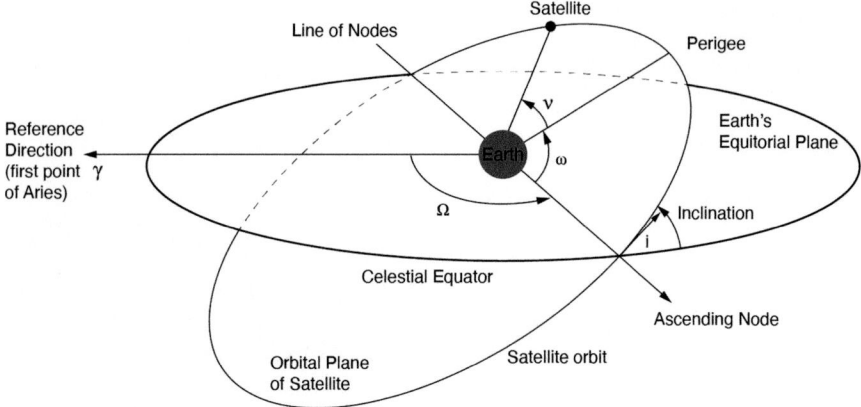

Fig. 5.6 The orbital elements of a satellite orbiting the Earth taking the Earth's equatorial plane as the reference plane. Note that the ascending node and descending node where the satellite crosses the reference plane. (Courtesy Katy Metcalf)

and 0° declination. Historically this point was in the constellation of Aries, but now due to the precession of the equinoxes, this point current lies in the constellation of Pisces.

Due to this drift of the equinoxes, a *reference epoch* is given, such as the current epoch J000.0 which represents noon, January 1, 2000 Terrestrial Time.[6] This fixes the reference direction of the first point of Aries for expressing celestial positions.

Besides inclination there are other important orbital elements that determine an orbiting body in space. Let us now switch the reference plane more suited to our local environment and use the Earth's *equatorial plane*, that is the plane extended into space by the Earth's equator, as our reference for a satellite orbiting the Earth although the orbital elements are applicable to any smaller body orbiting a larger one.

Unless the satellite is orbiting strictly over the Earth's equator, it will at some stage cross the equatorial plane. The point where the orbiting satellite passes through the reference plane from south to north is called the *ascending node* denoted by ☊, which looks like a set of headphones. Moreover, since we have a baseline direction, namely that given by the first point of Aries, it acts like a positive *x*-axis and the *longitude of ascending node*[7] is the angle from this reference direction to the ascending node of the orbiting satellite, denoted by Ω as in Fig. 5.6.

---

[6] Terrestrial Time (TT) is the international time standard adopted by the International Astronomical Union that is used for astronomical observations from Earth. It is used as the time for the positions of all Solar System bodies and forms the basis for Universal Coordinated Time (UTC).

[7] This is essentially the right ascension of the ascending node in this context.

Another orbital element is used in determining the current position of a satellite and that is the *argument of periapsis* which is the angle measured from direction of the ascending node extended to the direction of the *perigee* (in general periapsis) which lies along the semi-major axis. This determines the orientation of the elliptical orbit of the body in its orbital plane and the angle is denoted by ω.

Finally, there are a couple of anomalies. First is the *true anomaly* which is the angular displacement (denoted by ν) between the direction of perigee (periapsis) and the direction of the position of the satellite in its orbit at a particular time. This determines the position of the body with respect to its closest approach to the Earth. The second is the *mean anomaly*, which is the angle representing the fraction of a satellite's time since periapsis. Thus, if τ is the time of periapsis, and t is any other time, then the time elapsed since periapsis is $t - \tau$. The fraction of the entire orbit this elapsed time represents is $(t - \tau)/T$, where T is the orbital period. This is converted to an angular equivalent (in radians) by

$$M = \left(\frac{t-\tau}{T}\right) \times 2\pi,$$

which is the mean anomaly. Of course, 360° can be substituted by 2π. Thus the mean anomaly is a computational artifact.

Exercise Show that the mean anomaly M is the same as measuring the angle of a body from a given initial position with a circular orbit moving at constant velocity and having the same orbital period T as the original body. *Hint:* The angular velocity is defined by $\omega = \Delta\theta/\Delta t$, where $\Delta\theta$ is the angle displaced in the time period Δt.

Kepler's Laws

Although Kepler derived his three laws of planetary motion empirically, Isaac Newton deduced them mathematically from his Universal Law of Gravitation in his monumental *Principia*. This derivation was a revolutionary achievement for it meant that the workings of the Universe were now susceptible to mathematical consideration, an idea that was pursued vigorously in the centuries to follow. As mathematics is the language of terrestrial Physics, so it is also the language of the Heavens.

Kepler's 1st Law: All planets move around the Sun in elliptical orbits with the Sun at one focus (see Fig. 5.7).

Kepler's 2nd Law: All planets sweep out equal areas with the Sun in equal times (see Fig. 5.7). From this we can conclude that as the planet nears the Sun it must speed up.

Kepler's 3rd Law: The square of the orbital period of any planet is proportional to cube of its semi-major axis, and more specifically,

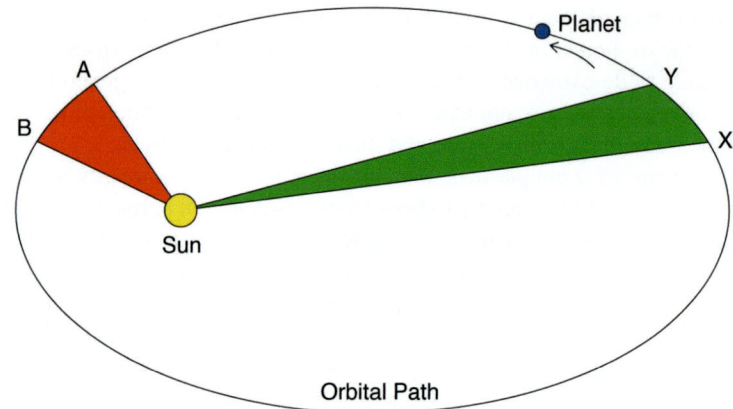

Fig. 5.7 A depiction of Kepler's 2nd Law whereby orbiting planets sweep out equal areas in equal time. This has the consequence that when planets are moving closer to the Sun their speed must increase. (Courtesy Katy Metcalf)

$$\boxed{T^2 = \left(\frac{4\pi^2}{GM_\odot}\right)a^3}. \quad (5.7)$$

This one is worthwhile to derive so let m be the mass of a body orbiting the Sun whose distance is r between their centers as in Fig. 5.8. We first need to discuss the notion of orbital velocity.

It is also now apparent why it is so important to have some in-depth knowledge of ellipses and their various components.

Orbital Velocity

Like Kepler's other two laws, this is a consequence of Newton's Law of Universal Gravitation, so let us just simplify matters (in spite of what the author just said above about ellipses) and take the case of a body of mass m in a *circular orbit* and derive Kepler's 3rd Law. The *centripetal force*[8] is that force required on the mass to maintain it in a circular motion with an orbital velocity v_o at a distance r from its center of rotation and directed inwards towards the axis of rotation as in Fig. 5.8 and is given by the formula

---

[8] The *centrifugal force* is a pseudo-force with the same magnitude but in the opposite direction to the centripetal force.

Kepler's Laws

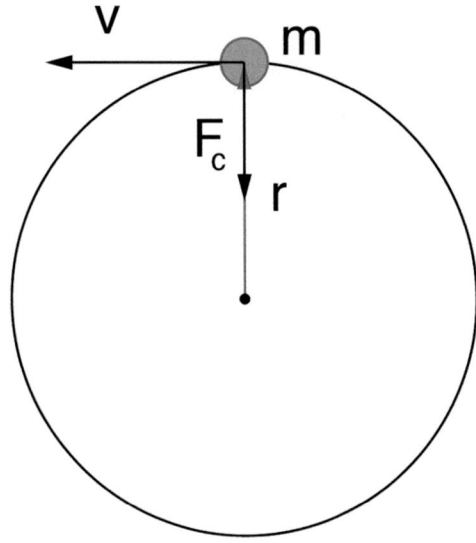

Fig. 5.8 The centripetal force F_c is that force required to maintain the body of mass m in a circular orbit around its central axis and is directed towards the axis of rotation. (Courtesy Katy Metcalf)

$$\boxed{F_c = \frac{mv_o^2}{r}}.$$

When the central point is the Sun, this centripetal force is just the force of gravity determined by the mass of the Sun $M_\odot$ acting on the mass m of the planet, namely

$$F = \frac{GM_\odot m}{r^2}.$$

Thus, we can set the two forces equal to obtain,

$$\frac{mv_o^2}{r} = \frac{GM_\odot m}{r^2},$$

implying that

$$v_o^2 = \frac{GM_\odot}{r}. \tag{5.8}$$

Applying the same reasoning more generally, note that for any small body of mass m in a circular orbit of radius r around a large mass M, the *orbital velocity* is therefore given by

$$\boxed{v_o = \sqrt{\frac{GM}{r}}}. \tag{5.9}$$

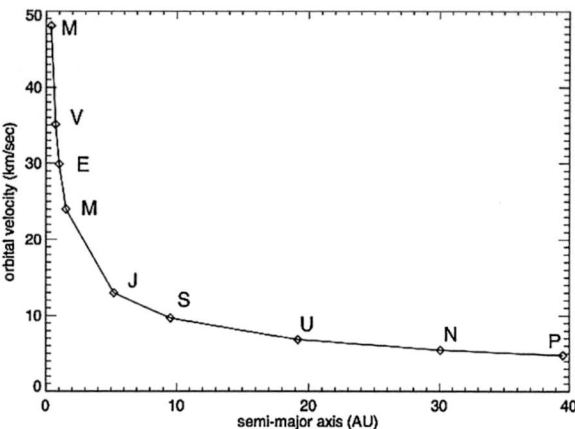

Fig. 5.9 The mean orbital velocity of the planets (including dwarf planet Pluto!) plotted as a function of the semi-major axis. Orbits were taken to be circular having circumference $2\pi a$. The orbital velocities exhibit the (Keplerian) behavior of $\frac{1}{\sqrt{a}}$. (Courtesy Chris Palma/Penn State)

This can also be expressed as

$$v_o \propto \frac{1}{\sqrt{r}},$$

and represents (Keplerian) *differential rotation* whereby bodies in outer regions are orbiting the Sun at a slower speed than bodies further in as in Fig. 5.9 for the orbital velocities of the planets (taking the semi-major axis a as a proxy for r).

This means that the more distant the object from the mass M, the lower its orbital velocity. For example, the mean orbital velocity of Earth is $v_o = 29.78$ km/s, whereas for Neptune it is $v_o = 5.45$ km/s. Moreover, when the Earth is at perihelion and hence slightly closer to the Sun in its orbit its velocity speeds up slightly to $v = 30.29$ km/s and at aphelion when slightly further away, its velocity is $v = 29.29$ km/s.

The Eq. (5.9) can be reformulated via the relation between arc length and displaced angle (in radians) for a circle: $s = r\theta$. Then

$$v_o = \frac{ds}{dt} = r\frac{d\theta}{dt} = r\omega,$$

where $\omega = d\theta/dt$ is the *angular velocity*. It follows from Eq. (5.9) that

$$\boxed{\omega = \sqrt{\frac{GM}{r^3}}}, \qquad (5.10)$$

and the centripetal force can be expressed as

$$\boxed{F_c = \frac{Mv_o^2}{r} = Mr\omega^2}. \qquad (5.11)$$

Kepler's Laws

Exercise Let M_1 and M_2 be the masses of two binary stars orbiting a common center of mass (the *barycenter*) at distances r_1 and r_2 respectively in isolated circumstances (a *two-body problem*). Then they are attracted to one another with a gravitational force

$$F = \frac{GM_1 M_2}{(r_1 + r_2)^2}.$$

Computing the centripetal force on M_1 and M_2 respectively in terms of angular velocity and equating them to F, show that the common angular velocity[9] of both masses is given by

$$\boxed{\omega = \sqrt{\frac{G(M_1 + M_2)}{(r_1 + r_2)^3}}}. \tag{5.12}$$

At this juncture, let us contrast Keplerian orbital velocity with *solid body rotation* of a point on a disk. In this instance every point on the (solid) disk will make a complete revolution in a time T (the period). A point at a distance r from the center will thus be moving at an orbital velocity of

$$v_o = \frac{2\pi r}{T}, \quad \left(\omega = \frac{2\pi}{T}\right),$$

and therefore, the orbital velocity of any point on the disk only depends in a linear manner on its distance r from the center, that is (see Fig. 5.10),

$$v_o \propto r.$$

(Note that the angular velocity is independent of the radius).

The difference found with differential (Keplerian) rotation is of course, gravity.

Exercise Calculate the mean orbital velocity of the Earth around the Sun from Eq. (5.9). *Ans.* 29.78 km/s.

---

[9] In a two-body problem, the common centripetal force experienced by each mass is

$$M_1 r_1 \omega_1^2 = \frac{GM_1 M_2}{(r_1 + r_2)^2} = M_2 r_2 \omega_2^2,$$

and in view of the 'center of mass formula' $M_1 r_1 = M_2 r_2$, these terms can be cancelled in the preceding formula, leaving $\omega_1^2 = \omega_2^2$ and so $\omega_1 = \omega_2$!

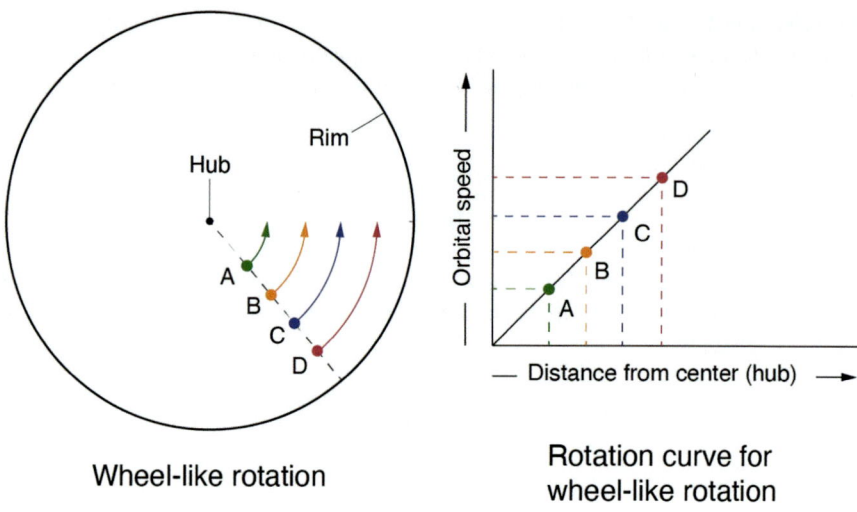

Fig. 5.10 Solid body rotation whereby every point on the solid disk has an orbital velocity that is linearly dependent on its distance from the center. (Courtesy Katy Metcalf)

Orbital Period/Mass

Back to Kepler. We next want to relate this result to the *orbital period/period of rotation* which is given by $T=$ distance/velocity $=2\pi r/v_o$, so that

$$T^2 = \frac{4\pi^2 r^2}{v_o^2} = \frac{4\pi^2 r^3}{GM_\odot}, \quad (5.13)$$

in view of Eq. (5.8). In general, for an extrasolar body of mass m orbiting a much larger star or black hole of mass M, we have in terms of the semi-major axis a of the smaller body, the slightly more general equation for an elliptical orbit

$$T^2 = \frac{4\pi^2 a^3}{G(M+m)}. \quad (5.14)$$

In simple terms this means that

$$T^2 \propto a^3.$$

Often the mass m is inconsequential compared to the mass M, and so we can use Kepler's Third Law to solve for the mass M if the period T is known. This gives us the equation,

Kepler's Laws

$$M = \frac{4\pi^2 a^3}{GT^2} \quad (5.15)$$

where a is the semi-major axis of the smaller orbiting body.

As in the case of the Sun and the Earth, the mass of the latter is insignificant as we have already determined the mass of the Earth from Newton's Law of gravitation.

Example (Mass of Sun) Let us compute the mass of the Sun using Eq. (5.15). For the semi-major axis, we will take (rounding off a bit):

$G = 6.6743$ m^3/kg · s^2;
$a = 149.598 \times 10^6$ km $= 1.496 \times 10^{11}$ m;
$a^3 = 3.3480 \times 10^{33}$ m^3;
$T = 365.25$ days $\times$ 86,400 s/day $= 3.156 \times 10^7$ s;[10]
$T^2 = 9.9603 \times 10^{14}$ s^2;
$4\pi^2 = 39.4784$.

Now we can put all these values together to obtain

$$M_\odot = \frac{4\pi^2 a^3}{GT^2} = \frac{(39.4784) \times (3.3480 \times 10^{33} \text{m}^3)}{(6.6743 \times 10^{-11} \text{m}^3/\text{kg} \cdot \text{s}^2) \times (9.9603 \times 10^{14} \text{s}^2)} = 1.988 \times 10^{30} \text{ kg}. \quad (5.16)$$

Exercise The relatively small innermost moon of Jupiter is Io, which has an orbital period of 1.769 days and its periapsis is 420,000 km and its apoapsis is 423,400 km. Compute the mass of Jupiter. *Ans.* 1.8988×10^{27} kg.

There is also a more user-friendly formulation of Kepler's Third Law that can be given. If the distance *a is given in terms of AU and the time T is given in terms of years*, then Eq. (5.15) can be expressed as[11]

---

[10] In dealing with Kepler's Laws and planetary dynamics, it is more accurate to use the sidereal period of $T_E = 365.25636$ days rather than the Julian period of 365.25 days, but in general, because we are rounding off to simplify matters, the difference is not significant.

[11] If a now represents the semi-major axis of the smaller body measured in terms of AU, and T represents its orbital period in units of years, then in standard units, $\bar{a} = \alpha a$, and $\bar{T} = \tau T$, with α, τ the units of distance in meters and time in seconds relative to the Sun, and thus $\bar{a}$ and $\bar{T}$ will be in units of meters and seconds respectively for the body of mass M. For this body of mass M by Eq. (5.15)

$$M = \frac{4\pi^2 \bar{a}^3}{G\bar{T}^2} = \frac{4\pi^2 (\alpha a)^3}{G(\tau T)^2} = \frac{a^3}{T^2}\left(\frac{4\pi^2 \alpha^3}{G\tau^2}\right) = \frac{a^3}{T^2} M_\odot$$

by Eq. (5.16).

Fig. 5.11 The asteroid designated as 1 Ceres; the first minor planet discovered on the first day of 1801 by the Catholic priest Giuseppe Piazzi. It is some 940 km in diameter and now enjoys the same elevated *dwarf planet* status as Pluto (which suffered a demotion in status) as well as others in our Solar System. The white spots on the surface are brine deposits arising from the interior. (Image taken by the Dawn spacecraft in 2015. Courtesy NASA/JPL-Caltech/UCLA /MPS/DLR/ IDA/Justin Cowart)

$$M = \frac{a^3}{T^2} M_\odot, \qquad (5.17)$$

again, with the proviso that the smaller mass m is insignificant compared to the larger mass M.

A nice example is presented below to compute the mass of Sagittarius A*. Likewise, if in Eq. (5.14) the period T is expressed in years and the average distance a is expressed in AU, we can write the equation in the simpler form

$$M + m = \frac{a^3}{T^2} M_\odot \qquad (5.18)$$

First however, let us note that if the orbited body *is the Sun* (so that $M = M_\odot$), then in view of Eq. (5.17) we obtain the simple formula

$$T^2 = a^3. \qquad (5.19)$$

Thus, for example, the period of the dwarf planet Ceres (Fig. 5.11) is $T = 4.6$ years so that its average distance from the Sun is

$$a = (4.6)^{2/3} = 2.766 \text{ AU}.$$

Conversely, we have seen previously in the chapter that the semi-major axis of comet Encke is $a = 2.2188$ AU. Therefore, its period is given by

Kepler's Laws

$$T = 2.2188^{3/2} = 3.3 \text{ years}.$$

Example Let us consider the binary star system Alpha and Beta Centauri.[12] Their orbital period is $T = 79.762$ yr and the distance to the pair is $D = 4.344$ ly with a common semi-major axis of $a = 17.493''$. We leave it as a small exercise for the reader to convert this angular value at the distance D to light-years and verify that $a = 3.684 \times 10^{-4}$ ly. Converting this to AU,

$$a = 3.684 \times 10^{-4} \text{ ly} \times \frac{1 \, AU}{1.58125 \times 10^{-5} \text{ly}} = 23.3 \text{ AU}.$$

Then by Eq. (5.18) we have

$$M_\alpha + M_\beta = \frac{23.3^3}{79.762^2} M_\odot = 1.988 \, M_\odot.$$

A breakdown of the two masses is: $M_\alpha = 1.0788 \, M_\odot$ and $M_\beta = 0.9092 \, M_\odot$.

*Mass of Sagittarius A**

Doing something more exotic we can also use Eq. (5.17) to find the mass of the black hole at the center of the Milky Way known as Sagittarius A* (Sgr A*).[13] This is done by examining the properties of nearby orbiting stars. Let us take a well-known star that has the designation S62 and has been observed for decades. In this instance, it has been found that the orbital period of S62 is:

$T = 9.9$ years so that $T^2 = 98.01$.

We require the semi-major axis in terms of AU. The semi-major axis has been measured in arcseconds θ_a:

$$\theta_a = 0.0905 \text{ arcsec} = 0.0905 \text{ arcsec} \times 4.8481 \times 10^{-6} \frac{\text{rad}}{\text{arcsec}} = 4.388 \times 10^{-7} \text{rad},$$

and we know that for such a small angle: $\tan \theta_a = \theta_a = 4.388 \times 10^{-7}$.

The distance to the black hole Sgr A* at the center of the Milky Way has been determined to be ~26,670 light-years. Thus, the distance in meters is[14]

---

[12] Actually, this is a triple star system with the third star being Proxima Centauri which is much further away from the alpha/beta system and has negligible effects on their dynamics.

[13] Black holes are discussed more comprehensively in Chap. 9.

[14] A reasonably accurate value here is essential in our determination as once it is multiplied times the angular displacement, it becomes cubed.

$$D = 26{,}670 \text{ ly} \times \left(9.46073 \times 10^{15} \text{m/ly}\right) = 2.523 \times 10^{20} \text{m}.$$

Since we know that

$$\tan \theta_a = \frac{a}{D},$$

we have $a = D \tan \theta_a = (2.523 \times 10^{20} \text{m}) \times (4.388 \times 10^{-7}) = 11.071 \times 10^{13}$ m, which converts the semi-major axis to meters. In terms of AU we obtain

$$a = \frac{11.071 \times 10^{13} \text{ m}}{1.496 \times 10^{11} \text{ m/AU}} = 7.403 \times 10^2 \text{ AU}.$$

Therefore, $a^3 = 405.72 \times 10^6$.

This is all we need. Then computing the mass via Eq. (5.17)

$$M = \frac{a^3}{T^2} M_\odot = \frac{405.72 \times 10^6}{98.01} = 4.14 \times 10^6 M_\odot.$$

We have thus found that the mass of the black hole Sgr A* is

$$\boldsymbol{M = 4.14 \times 10^6 M_\odot},$$

or about 4 million times the mass of our Sun.

There are various other stars that have been used for this calculation (see Fig. 5.12) and the average can be taken. For example, the star S4711, announced in 2020[15] has a period of $T = 7.6$ yr and apoapsis $r_a = 1094.7$ AU, periapsis $r_p = 143.7$ AU. Then the semi-major axis is given by their average (Eq. (5.3))

$$a = \frac{r_a + r_p}{2} = 619.2 \text{ AU}.$$

Then the mass of Sgr A* is found to be

$$M = \frac{a^3}{T^2} M_\odot = \frac{(619.2)^3}{(7.6)^2} M_\odot = 4.11 \times 10^6 \ M_\odot.$$

It should be mentioned that there is a certain amount of error margin attached to the apoapsis and periapsis so this value should be understood within that context.

---

[15] F. Peißker et al., S62 and S4711: Indications of a Population of Faint Fast-moving Stars inside the S2 Orbit—S4711 on a 7.6 yr. Orbit around Sgr A*, *A J.*, 899:50 (2020), 19 pp.

Kepler's Laws

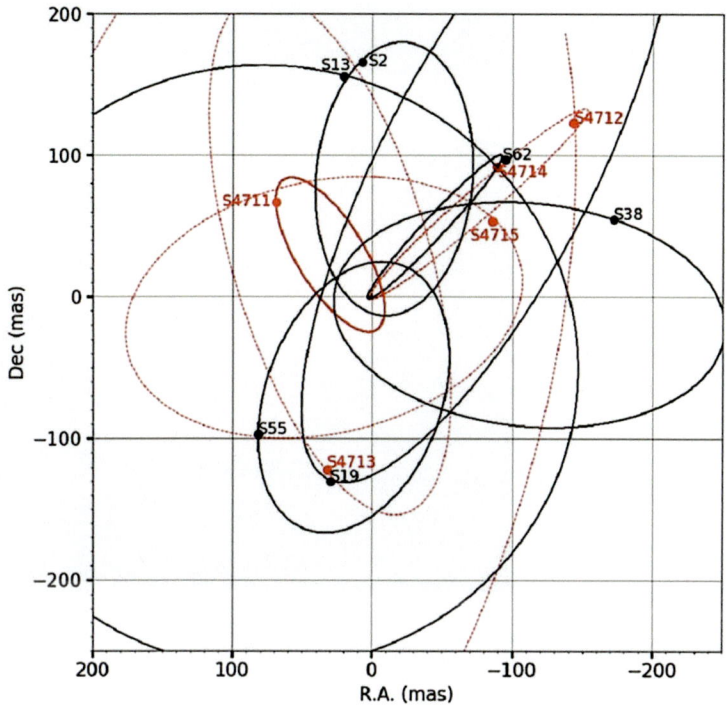

Fig. 5.12 Image showing the orbits of several S-stars orbiting Sgr A* which is located at the origin of the figure. Newly imaged star orbits in red. The orbit of S4711 is in solid red and S62 in black. The star S4714 has been identified as having the fastest known maximum velocity at 24,000 km/s. (Courtesy F. Peißker et al.)

A commonly accepted value for the mass of the black hole Sgr A* is

$$M_{SgrA*} = 4.15 \times 10^6 \, M_\odot,$$

so let us accept this value for any future calculations. See Fig. 5.13.

For their work on Sgr A*, astronomers Andrea Ghez and Reinhard Genzel shared half the 2020 Nobel Prize in Physics, with the other half going to physicist Roger Penrose for his theoretical work on black holes.

Exercise* (referred to later in the text). The star S2 (S0–2) orbits Sgr A* with a semi-major axis of $a = 0.12540$ arcsec and an orbital period of $T = 16.0518$ years.

(a) Compute the length of the semi-major axis a in meters and then in AU. *Ans.* 15.345×10^{13} m; 1025.7 AU.
(b) Compute the mass of the black hole Sgr A* from this data. *Ans.* $4.19 \times 10^6 \, M_\odot$.

Fig. 5.13 Image of the black hole at the heart of our Milky Way, Sgr A*. The bright ring primarily represents the glowing hot dense disk of gas and matter spiraling into the black hole of the *accretion disk*. Some of this light is also due to the light from background objects following curved paths due to the warping of spacetime near the black hole. (Courtesy Event Horizon Telescope 2022)

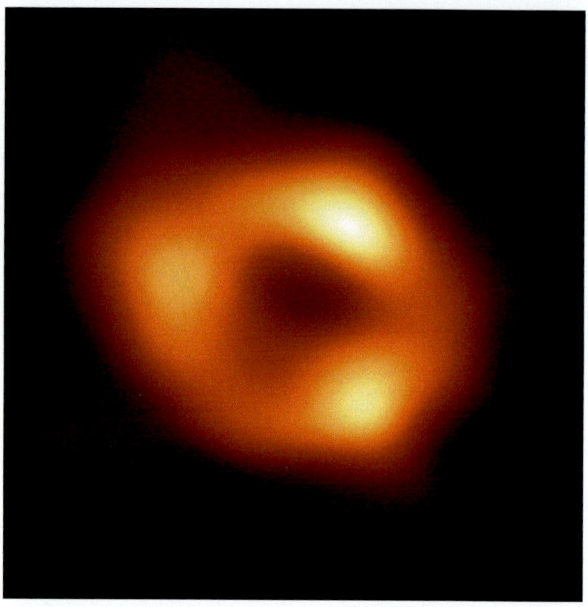

Another relationship that can be gleaned from Kepler's 3rd Law is to consider two planets orbiting the Sun, say of semi-major axes a_1 and a_2 and having orbital periods T_1 and T_2 respectively. It follows that since the mass of the Sun is the same for both cases,

$$\frac{4\pi^2 a_1^3}{GT_1^2} = \frac{4\pi^2 a_2^3}{GT_2^2}$$

and rearranging the terms

$$\boxed{\frac{T_1^2}{T_2^2} = \frac{a_1^3}{a_2^3}}. \quad (5.20)$$

This means that if we know any three of the terms, we can find the fourth term. For example, if we know that the period of Jupiter say, is $T_J = 4332.6$ days, we can use that to find its semi-major axis using the fact that for the Earth, $T_E = 365.25$ days and $a_E = 149.6 \times 10^6$ km. From Eq. (5.20) we can work out *Jupiter's semi-major axis a_J*, that is,

$$a_J^3 = \frac{T_J^2 \times a_E^3}{T_E^2} = \frac{(4332.6)^2 \times (149.6 \times 10^6)^3}{(365.25)^2} = \cdots$$

Kepler's Laws

The rest of the calculation will be left to the reader. Depending on rounding off, the value for the semi-major axis of Jupiter should be $a_J \approx 778 \times 10^6$ km having taken the cube root of the right-hand side.

Geostationary Satellites

Satellites used for communication and weather purposes are in a geostationary orbit remaining in the same spot above the Earth and we assume that the satellite's circular orbit lies in a plane through the Earth's equator. Thus, the period of the satellite's orbit must match that of the Earth that is, $T = 24$ h $= 86,400$ s $= 8.640 \times 10^4$ s.

Since we know the mass of the Earth is $M_E = 5.972 \times 10^{24}$ kg, we can find the radius of the orbit of the stationary satellite from Eq. (5.15) and solving for the radius a, that is,

$$a^3 = \frac{M_E G T^2}{4\pi^2} = \frac{(5.972 \times 10^{24} \text{ kg}) \times (6.674 \times 10^{-11} \text{m}^3/\text{kg} \cdot \text{s}^2)(8.640 \times 10^4 \text{ s})^2}{39.48}$$

$$= 75.36 \times 10^{21} \text{ m}^3.$$

Therefore,

$$a = 4.2239 \times 10^7 \text{ m} = 42,239 \text{ km}.$$

However, this is the distance *from the center of the Earth* and so the altitude above the Earth's surface is given by

$$h = 42,239 \text{ km} - 6371 \text{ km} = 35,868 \text{ km}.$$

What about the velocity of the geostationary satellite? For this we simply require Eq. (5.9) in the form

$$v_o = \sqrt{\frac{GM_E}{a}},$$

and solve as in the preceding example. We leave it to the reader to verify that $v \approx 3$ km/s.

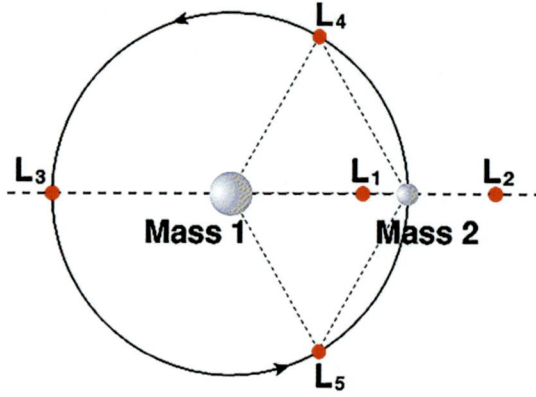

Fig. 5.14 The five stationary Lagrange points determined by the Sun and Earth. (Image public domain)

Lagrange Points

If we consider just two celestial bodies orbiting about a common center of mass, it turns out that there are actually five different points where a small body placed at any of these points would remain stationary. The first three colinear points $L1$, $L2$, $L3$ were found by the mathematician Leonhard Euler and the latter two by Joseph-Louis Lagrange. The labelling can vary and the order of the colinear points $L2$, $L1$, $L3$, is also used by NASA. Note that two equilateral triangles are formed by the two celestial bodies and $L4$ or $L5$ is in Fig. 5.14.

In the Sun/Earth system, any satellite inside the Earth's orbit will have a velocity faster than that of Earth (Eq. (5.9)), but if the satellite is just close enough to the Earth at $L1$, the Earth's gravity will allow the satellite to slow down and match the Earth's orbital velocity. This is the location of the Solar and Heliospheric Observatory (SOHO) satellite observing the Sun about 1.5 million kilometers from Earth. Assuming a circular orbit for the Earth and if a is the distance between the Sun and Earth, then the position of $L1$ as a distance from M_2 is given by

$$\boxed{L1_D \approx a \left(\sqrt[3]{\frac{M_2}{3(M_1 + M_2)}} \right)}, \tag{5.21}$$

but we will not pursue the derivation as it is a bit involved.[16] But again, as in the case of one body orbiting another, it is a matter of the gravitational forces of the two massive bodies matching the centripetal force experienced by the small orbiting

---

[16]This distance corresponds to the radius of the 'Hill sphere of influence' which is the spherical region around a celestial body where its gravitational force dominates over the gravitational force of a more massive body that it orbits as in the case of a planet orbiting the Sun. Within this sphere of influence the body is able to retain satellites in a stable orbit.

body. The $L1$, $L2$, $L2$, Lagrange points are considered unstable so that satellites placed at any of these points require regular course corrections.

However, let us calculate the position of $L1$ from the preceding formula. Taking:

$a = 1.496 \times 10^8$ km
$M_1 = M_\odot = 1.989 \times 10^{30}$ kg
$M_2 = M_E = 5.972 \times 10^{24}$ kg

then the distance of the $L1$ point from the Earth in view of Eq. (5.21) is given by

$$L1_D \approx \left(1.496 \times 10^8 \text{km}\right) \left(\sqrt[3]{\frac{5.972 \; 10^{24} \text{ kg}}{3 \times \left(1.989 \times 10^{30} \text{ kg}\right)}} \right) = 1.496 \times 10^6 \text{ km},$$

where in this instance the role of the mass of the Earth is not significant in our calculation. Interestingly, note that the distance to the $L1$ point is ~100th the distance of the Earth to the Sun.

On the other hand, in the Sun/Earth system, any satellite outside the Earth's orbit will have a slower orbital velocity than the Earth's, but at just the right spot, namely $L2$, where the Earth's and Sun's gravitational attraction combine, the satellite will have the same orbital velocity as the Earth and remain in a stationary position with respect to the Earth. This is the current position of the James Webb Space Telescope at a distance of ~1.5 million kilometers on the opposite side of the Earth from the SOHO satellite and again ~100th the Earth-Sun distance.

In the case of the Sun and the planet Jupiter, there are two sets of asteroids, each orbiting at the Lagrange points $L4$ and $L5$. The Trojans are a group located at the $L4$ point 60° ahead of Jupiter and at the $L5$ point 60° trailing the planet. Similarly, Mars, Neptune, and Uranus also have a group of Trojans at the $L4$ and $L5$ points and even the Earth has two Trojan asteroids, both at the $L4$ Lagrange point. Venus has a temporary Trojan as does the dwarf planet Ceres and the asteroid Vesta.

Deviations from Kepler's Laws Reveal Dark Matter

As is the case with planets and satellites in our Solar System, the Newtonian dynamics of a spiral galaxy dictate that stars increasingly distant from the center of the host galaxy should have a slower orbital velocity. Specifically for a star at a radius R from the center of the galaxy, its orbital velocity will be given by Eq. (5.9) in the form

$$\boxed{v_o(R) = \sqrt{\frac{G \cdot M(R)}{R}}}, \tag{5.22}$$

where $M(R)$ is the mass of the galaxy inside the radius R.[17] For spiral galaxies, in the core region of the central bulge of the galaxy, let us assume that the bulge is spherical so that the mass can be modeled by

$$M(R) = \rho \cdot \frac{4}{3}\pi R^3,$$

assuming that the density of the bulge ρ is constant. Then

$$v_o(R) = \sqrt{\frac{G\rho\left(\frac{4}{3}\right)\pi R^3}{R}},$$

which means that

$$v_o(R) \propto R,$$

which is just a linear relation. Outside the central bulge, again assuming a constant density (but using the same symbol) in a flattened disk,

$$M(R) = \rho \cdot \pi R^2,$$

and consequently, in this disk region

$$v_o(R) = \sqrt{\frac{G\rho\pi R^2}{R}},$$

and,

$$v_o(R) \propto \sqrt{R}.$$

Furthermore, at a distance R_0 beyond the core region *and* the luminous disk component, observe that in Fig. 5.15 the orbital velocities are essentially constant with increasing distance from the center. For this to happen, then in view of Eq. (5.22)

$$\sqrt{\frac{G \cdot M(R)}{R}} = \text{constant},$$

which is to say that we require

---

[17] As with the planets there is a small additional contribution to the orbital velocity from the mass outside the star's orbit but this will be ignored in this instance.

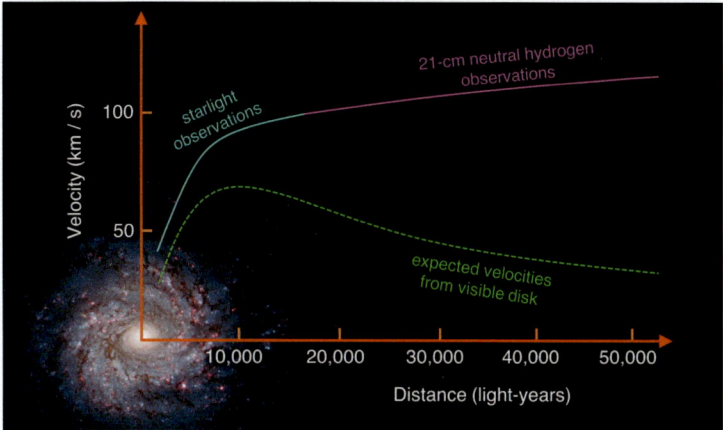

Fig. 5.15 A typical rotational velocity curve obtained for a spiral galaxy showing both the measured velocities via 21-cm neutral hydrogen (see Chap. 7) and the expected rotational velocities via Newtonian dynamics (values are for illustration purposes only). Notice the flattening out of the measured orbital velocities for stars in the outer regions of the disk beyond 15,000 ly where there is little galactic mass. (Image of galaxy NGC 3982 courtesy NASA, ESA, and the Hubble Heritage Team (STScI/AURA), and Jonathan Park for graphics overlay)

$$\boxed{M(R) \propto R} \qquad (5.23)$$

in this outer region. This means that the mass of the galaxy must continue to increase so that doubling the distance from say 10 kly to 20 kly should double the mass inside this radius.

On the other hand, the traditional scenario is that beyond the central bulge and the luminous disk component, it is assumed that $M(R) \approx$ constant which should then result in Keplerian rotation in the very outer reaches of the galaxy. That is, just as in the case of planets orbiting the Sun, for stars or gas clouds in the galaxy's outer regions their orbital velocity in view of Eq. (5.22) should diminish as

$$v_o(R) \propto \frac{1}{\sqrt{R}},$$

again where $R > R_0$ is the distance from the galactic center as indicated by the decreasing portion of the dashed line in Fig. 5.15. Since this is not the case, in this traditional view the mass required by Eq. (5.23) is not observed and thus, there must be some unseen mass to account for the flat orbital velocity curves in the outer regions of the galaxy. This unseen matter is what is termed *dark matter* and discussed at length below.

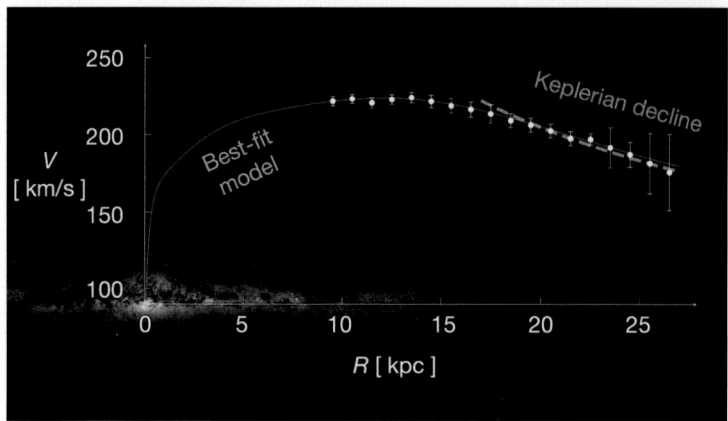

Fig. 5.16 The velocity rotation curve for the Milky Way galaxy indicating the orbital speed of stars as a function of distance from the galactic center. The measurements were taken from the *Gaia* DR3 data catalogue. The curve indicates the best fit model that includes both ordinary and dark matter. The dashed yellow line indicates a Keplerian decrease as $R^{-1/2}$ from beyond the optical part of the disk. This indicates much less dark matter than expected. (Courtesy Jiao, Hammer et al./ Observatoire de Paris—PSL/CNRS/ ESA/Gaia/ESO/S. Brunier)

Interestingly, recent research based on the *Gaia* GR3 dataset[18] on the orbital velocities for stars in the Milky Way apparently does indicate a Keplerian $R^{-1/2}$ decline (Fig. 5.16), suggesting that either our galaxy is in some ways exceptional regarding its structure/history, or due to differences in methodology that was employed in this as well as other similar studies. Or perhaps, it is just a dip in the road that will disappear at further distances. As a consequence of the *Gaia* data, the mass of the Milky Way works out to be ~2 × 10^{11} solar masses (see Exercise that follows), well below previous values 10^{12} solar masses.

Indeed, for a very simplified calculation of the mass of the Milky Way, let us consider the orbiting Large Magellanic Cloud (LMC) at a distance of 50 kpc and a mean orbital velocity of 300 km/s. Rearranging the terms in Eq. (5.22) we have

$$M(R) = \frac{R v_o^2(R)}{G}.$$

Then,

$$M(R) = \frac{(5 \times 10^4 \text{ pc}) \times (3.09 \times 10^{16} \text{ m/pc}) \times (3 \times 10^5 \text{ m/s})^2}{6.67 \times 10^{-11} \text{ m}^3/\text{kg} \cdot \text{s}^2} = 2.08 \times 10^{42} \text{ kg}.$$

---

[18] See: Y Jiao et al., Detection of the Keplerian decline in the Milky Way rotation curve, *A&A*, 678 (2023), 13 pp.

Dividing this value by the solar mass: $M_\odot = 1.99 \times 10^{30}$ kg, gives a basic estimate of the mass of the Milky Way as **~1.05 × 10¹² $M_\odot$**. The value of 10^{12} solar masses has also been determined by other means. So, the value derived from the *Gaia* data has been quite disruptive and further indicates a much lower dark matter content surrounding our galaxy than had been previously thought.

This tension between the behavior seen in Figs. 5.15 and 5.16 and the mass of the Milky Way discrepancy is as of the present date (2024) driving a lot of astronomical research. This is real Astronomy in the making and a very good lesson in how Science works.

Exercise Do a back of the envelop calculation from Fig. 5.16 by taking the radius of the Milky Way as 17 kpc and the orbital velocity of a star at that distance to be 220 km/s. Using Eq. (5.22), show that the mass of the Milky Way is $\sim 2 \times 10^{11}\ M_\odot$.

Going back to the flat rotation curves arising from $M(R) \propto R$ in the outer regions of a spiral galaxy, then the density of the matter there would be

$$\rho(R) = \frac{M(R)}{V(R)} \propto \frac{R}{R^3} = \frac{1}{R^2}.$$

Indeed, computer simulations have led to a density profile of the *dark matter halo* given by

$$\boxed{\rho(R) = \frac{k}{R(a+R)^2}}, \qquad (5.24)$$

where k and a are constants. Note that as $R \to 0$, the density becomes arbitrarily large indicating a high density near the core and in the outer regions of the galaxy the density falls off as R^{-3}.

Of course, dark matter still remains a contentious notion as various mechanisms for its existence have been proposed, such as WIMPS (Weakly Interacting Massive Particles), Axions (lighter than WIMPS), MACHOs (Massive Compact Halo Objects), Dark Atoms, etc. but which remain speculative and none have been found.

Other approaches are more theoretical in an attempt to explain the dynamics required by DM. One of the best known is *modified Newtonian dynamics* (MOND) due to Israeli physicist Mordechai Milgrom which is a modification of the laws of gravity that can also account for the flat orbital velocity curves seen in spiral galaxies and does not have to invoke dark matter. However, it does have certain issues when encountering the evidence from the collision between two clusters of galaxies.[19]

Indeed, such is the case in Fig. 5.17 where the hot gas in the two clusters collided at very high velocity and subsequently slowed down. On the other hand, the dark matter passed right through the collision and resides in the blue regions of the figure

---

[19] See book by the author, *The Most Interesting Galaxies in the Universe* in Bibliography.

Fig. 5.17 The cluster MACS J0025.4-1222 depicting a collision of two galaxy clusters and indicates how the dark matter became separated from the ordinary matter. Most of the latter is found in the superheated glowing intracluster gas, in pink from X-rays, whereas most the matter of the entire cluster is in the blue regions which consists of ordinary and dark matter. (Courtesy NASA, ESA, CXC, M. Bradac (University of California, Santa Barbara), and S. Allen (Stanford University))

where the majority of the mass lies. The glowing pink regions are the hot gas detected by the Chandra X-ray telescope and is where the majority of ordinary matter resides. This unusual distribution of ordinary and dark matter was achieved through gravitational lensing discussed in Chap. 8 and is another confirmation of the existence of dark matter. A similar effect has been found in other collisions of galaxy clusters such as the Bullet Cluster among others.

As well there are other theories such as that of A. Sipols and A. Pavlovich[20] where the authors claim to show that the disk structure and peripheral mass exterior to where the orbit measurements are taken makes a sufficient difference to the rotation curves to explain their flat nature without the need to invoke dark matter or theories of modified gravity.

Another theory that has found a recent new impetus is that of *primordial black holes* (*PBH*) which were proposed in the work of Russian physicists Y. B. Zel'dovich and I. D. Novikov in 1966 and further expounded upon by S. Hawking. Unlike standard black holes which formed as the Universe evolved over eons, PBHs are thought to have formed in the early Universe after the Big Bang and can vary in mass from 10^{-5} g to thousands of times the mass of the Sun. They are thought to be a consequence of small density fluctuations in the primordial soup that followed the Big Bang. The size of the black hole is governed by the Schwarzschild radius discussed in Chap. 9. For example, a black hole having the mass of the Moon will have a radius of 0.1 mm. While no primordial black holes

---

[20] Dark Matter Dogma: A Study of 214 Galaxies, *Galaxy*, 8 (2020) 32 pp.

Fig. 5.18 The Coma Cluster consisting of more than 1000 galaxies some 320 light-years distant. It was the anomalous velocities of its galaxies as determined by Fritz Zwicky that gave the first indication of dark matter. The circular arcs are the distorted images of galaxies well behind the cluster due to gravitational lensing caused by the ordinary and dark matter in the cluster. The visible matter alone is insufficient to cause the warping of spacetime resulting in the arcs. See Chap. 8. (Courtesy NASA, ESA, J. Mack (STScI) and J. Madrid (Australian Telescope National Facility))

have yet been detected, they remain a strong contender for an explanation of dark matter and an active area of current research.

And so on and so on... The dust has not yet settled on which approach is the correct one.

Exercise Assuming a spherical mass distribution of the Milky Way (of course, this is not actually the case), and the Sun's orbital velocity of 230 km/s at a distance of 26,670 light-years from the center, compute the mass $M(R)$ of the Milky Way interior to the Sun's orbit via Eq. (5.22). *Ans.* $\approx 2 \times 10^{41}$ kg $\approx 1 \times 10^{11} M_\odot$.

The missing-mass anomaly occurs elsewhere in the Universe and has an even earlier pre-history. In fact, it was the Swiss-American astronomer Fritz Zwicky[21] who in the 1930s was observing the velocities of the galaxies in the Coma Cluster (Fig. 5.18) and used the Virial Theorem Eq. (4.6) to determine the total mass of the cluster. However, the mass calculated from the high velocities of the galaxies was significantly greater than what could be accounted for by the visible luminous matter.

The calculation is illuminating (not only to see a genius mind at work) and we reproduce here exactly what Zwicky calculated using his original notation and the

---

[21] Fritz Zwicky (1898–1974) worked most of his life at the California Institute of Technology in Pasadena and is remembered as both a genius and rather cantankerous.

same steps of his calculation so that we can see history being made first hand.[22] The whole procedure can be simplified in Exercise (a) to follow. Zwicky took the Coma Cluster to be composed of 800 galaxies each of which had a mass of roughly $10^9 M_\odot$ so this gives a total mass of

$$M = 800 \times 10^9 \times 2 \times 10^{30} \text{ kg} = 1.6 \times 10^{42} \text{ kg}.$$

The total potential energy is given by Eq. (4.5), namely

$$U = -\frac{3}{5}\frac{GM^2}{R},$$

where G is the gravitational constant and R is the radius of the cluster which Zwicky took to be 1 million light-years, that is,

$$R = 10^6 \text{ ly} \times 10^{16} \frac{\text{m}}{\text{ly}} = 10^{22} \text{ m}.$$

Therefore, the *mean potential energy per unit mass* of the system is given by

$$\bar{\varepsilon}_P = \frac{U}{M} = -\frac{3}{5}\frac{GM}{R},$$

so that doing the calculation gives

$$\bar{\varepsilon}_P = -\frac{3}{5}\frac{\times \left(6.67 \times 10^{-11} \frac{\text{m}^3}{\text{kg}\cdot\text{s}^2}\right) \times 1.6 \times 10^{42} \text{ kg}}{10^{22} \text{ m}} = -6.4 \times 10^9 \text{ m}^2/\text{s}^2.$$

On the other hand, the mean kinetic energy $(\frac{1}{2}M\bar{v}^2)$ per unit of mass is

$$\bar{\varepsilon}_K = \frac{1}{2}\bar{v}^2,$$

and applying the Virial Theorem Eq. (4.6) to $\bar{\varepsilon}_K$ and $\bar{\varepsilon}_P$ we obtain

$$\frac{1}{2}\bar{v}^2 = \bar{\varepsilon}_K = -\frac{1}{2}\bar{\varepsilon}_P = 32 \times 10^8 \text{ m}^2/\text{s}^2.$$

Solving for the mean velocity gives

---

[22] Die Rotverschiebung von extragalaktischen Nebeln, *Helv. Phys. Acta* 6 (1933) 110–127. In those days, galaxies were still referred to as 'nebulae'. Zwicky measured distance in centimeters whereas we have used meters but that is the only difference in our calculations.

$$\bar{v} = 8 \times 10^4 \text{ m/s} = 80 \text{ km/s}.$$

This very low mean velocity value resulting from the total mass M of luminous matter does not square at all with the differences in the velocity of some of the galaxies in the Coma Cluster on the order of 1500 – 2000 km/s.[23] This was the dilemma faced by Zwicky.

The upshot is that there had to be some unaccounted-for mass that he called *dunkle Materie*, that is, dark matter, since it emits no observable light. Of course, we now know that there is also intergalactic gas weighing more than the luminous stars but this is still insufficient to account for the galaxy velocities and dark matter remains the dominant mass of the cluster.

Unfortunately, Zwicky's dark matter hypothesis did not receive very much attention until the 1970s when missing mass showed up in various scenarios such as in the case of spiral galaxies. Now most astronomers accept the existence of dark matter as a tenable notion to explain the anomalous orbital velocity behavior due to missing observable matter. This would take the form of a spherical halo of unseen mass surrounding the galaxy. It is called *cold* dark matter as it is believed that the responsible particles move much slower when compared to the speed of light. Some 27% of the Universe is thought to consist of dark matter, with ordinary matter comprising only about 5%. See Chap. 10 for a more complete discussion of the composition of the Universe.

Exercise

(a) Using the Virial Theorem, show that the average velocity $\bar{v}$ of a particle in the system can be written as

$$\bar{v} = \sqrt{\frac{3}{5}\frac{GM}{R}}.$$

(b) A recent study has measured the mass of the Coma Cluster (including dark matter) as 1.8×10^{15} solar masses and using the radius of 3.067 Mpc, compute the average galaxy velocity $\bar{v}$. *Ans.* 1231 km/s.

Regarding the dark matter of the Coma Cluster, it also makes itself evident in the gravitational lensing effect it has on distant galaxies behind the cluster as will be discussed in Chap. 8.

---

[23] The individual motions of the galaxies in the cluster are influenced by the gravitational pull of other galaxies in the cluster, which leads to a range of velocities. See the Exercise (b) in this section.

Chapter 6
Climbing the Distance Ladder

Distance Formula

In dealing with distance, there is one basic formula that is most essential, namely

$$\boxed{D = v \cdot t} \tag{6.1}$$

that was discussed in Chap. 1 regarding light-years. Let us do something with it more interesting regarding the pair of interacting galaxies Arp 147 (Fig. 6.1). These represent the aftermath of an interaction of two galaxies with the one on the left having passed through the one on the right leaving the circular ring of intense star formation remnant from a former spiral galaxy and a tidal ring remnant around the intruder elliptical galaxy.

The radius of the blue ring has been measured at $R = 5.8$ kpc. Since $1\text{pc} = 3.09 \times 10^{13}$ km, the radius is

$$R = (5.8 \times 10^3 \text{ pc}) \times \left(3.09 \times 10^{13} \frac{\text{km}}{\text{pc}}\right) = 1.79 \times 10^{17} \text{ km}.$$

Moreover, the expansion of the outer diameter of the blue ring has been measured at a velocity of $v = 225$ km/s. This means that matter in the outer ring has been travelling from the center of the ring at a velocity of *half* of that, that is, 112.5 km/s. Therefore, taking the radius as our travel distance as $D = R = 1.79 \times 10^{17}$ km, then matter on the outer edge of the blue ring moving at a rate of $v = 112.5$ km/s,[1] has been travelling for a time period of

---
[1] We will assume that the velocity has been constant since the time of collision.

Fig. 6.1 Arp 147 showing the aftermath of the collision of two galaxies with the one on the left having passed through the one on the right leaving a tidal ring around the elliptical intruder galaxy and the circular ring remnant of a former disk galaxy on the right. (Courtesy NASA/HST)

$$t = \frac{D}{v}$$
$$= \frac{1.79 \times 10^{17} \text{ km}}{112.5 \text{ km/s}} = 159 \times 10^{13} \text{ s}.$$

Converting this to years since 1 year equals 3.16×10^7 s gives $\approx$50 million years, which is the approximate time in the past when the collision took place. And with this we have done some real Astronomy.

Exercise The mean radius of the Earth is $r = 6371$ km. Taking the period of rotation as 24 h, what is the Earth's rotational velocity at the equator in km/s? *Ans.* 0.463 km/s.

Parallax

Anyone who has ever stuck their finger in front of their face and observed how their finger seems to move back and forth when you alternately open and close your left and right eye will be familiar with the concept of parallax. If we replace your finger with a stellar object such as a nearby star, and replace your nose with the Sun, and each eye by a telescope on the surface of the Earth at opposite points in its orbit, we can play the same game but on a larger scale. Nearby stars will appear to move against their celestial background when the Earth is on opposite sides of the Sun during the year as depicted in Fig. 6.2. The parallax is the angle α of the figure and it is a simple matter of trigonometry to determine its value.

Parallax

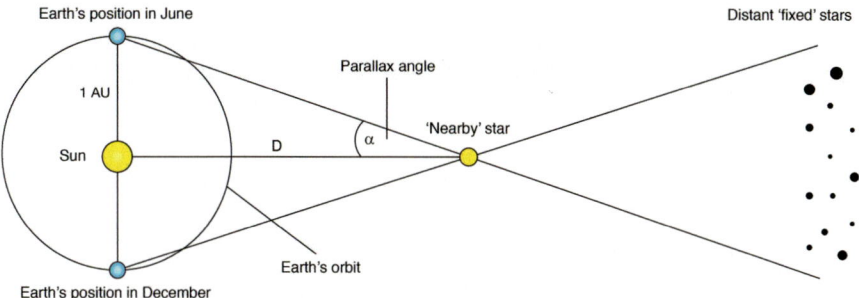

Fig. 6.2 The position of a nearby star is measured from positions when the Earth is 6 months apart as in the figure. (Credit: Katy Metcalf)

Consider a nearby star, the red dwarf Barnard's Star, whose parallax angle is $\alpha=$ 0.547 s of arc. The radius of the Earth's orbit is 149.6 million km, but as mentioned previously this is just an average value as the orbit is somewhat elliptical and not a perfect circle. From the diagram,

$$\frac{\text{radius}}{\text{distance}} = \tan \alpha \quad \text{i.e., distance} = \frac{\text{radius}}{\tan \alpha}.$$

To carry out the calculation let us convert the units of arcseconds to radians[2] and in this case, 0.547 arcsec $= 2.652 \times 10^{-6}$ radians. Therefore, $\tan\alpha = 2.652 \times 10^{-6}$, and now observe something particularly interesting, that as in this case *the tangent of a very small angle has the same numerical value as the angle itself in radians* up to several decimal places). This fact is known as the *small angle approximation* and holds for the sine function as well, both up to about 0.2 radians.[3] So, we have for small angles using radians

$$\tan \alpha = \alpha,$$

implying

$$\boxed{\text{Distance} = \frac{\text{radius}}{\alpha}}. \tag{6.2}$$

The distance D becomes:

---

[2] We know from Chap. 1 that 1 arcsec $=\pi/648{,}000$ rad.
[3] This fact may not be taught in a course on Trigonometry but it is known by all astronomers because they are routinely dealing with small angles.

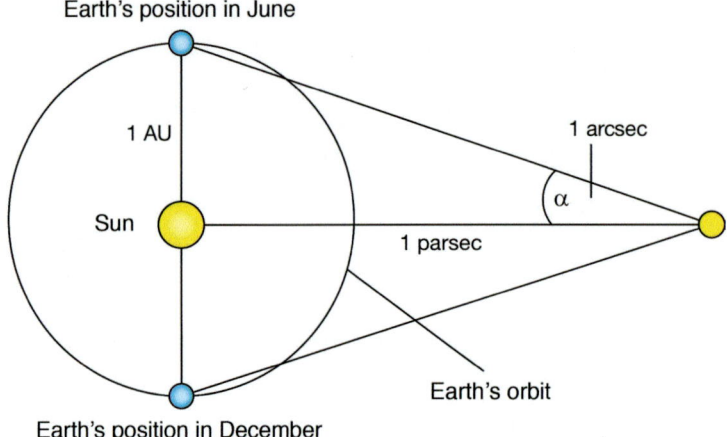

Fig. 6.3 A parsec is the distance to an object that subtends an angle of 1 arcsec with the 1 AU radius of the Earth's orbit. (Courtesy Katy Metcalf)

$$D = \frac{149.6 \times 10^6 \text{ km}}{2.652 \times 10^{-6}} = 56.410 \times 10^{12} \text{ km}.$$

This value has little meaning, but we know that there are 9.462×10^{12} km in a light-year (ly), which gives us the distance to Barnard's Star as[4]

$$D = \frac{56.410 \times 10^{12} \text{ km}}{9.462 \times 10^{12} \text{ km/ly}} = 5.96 \text{ ly}.$$

But this was a rather laborious method for computing parallax and we can avoid radians altogether with a much simpler calculation involving the notion of a *parsec* (pc). We will define a *parsec* as the distance to an object when the 1 AU radius of the Earth's orbit subtends an angle of 1 arcsec as in Fig. 6.3. Recalling that 1 arcsec=π/648,000 rad, we define

$$\mathbf{1\ parsec} = \frac{1 \text{ AU}}{\tan \alpha} = \frac{1 \text{ AU}}{\alpha} = \frac{648{,}000 \text{ AU}}{\pi} = \mathbf{206{,}265\ AU}.$$

And if we put in the value of 1 AU $=1.58125 \times 10^{-5}$ ly we find that

$$1 \text{ parsec} = 206{,}265 \text{ AU} \times 1.58125 \times 10^{-5} \text{ ly/AU}$$

that is,

---

[4]Due to rounding off in various calculations there will always be some slight discrepancy in published figures. But due to the large values involved, this is of little consequence.

$$1\text{ pc} = 3.2616 \text{ ly},$$

which is often just rounded off to **1pc = 3.26 ly**.

In general, if we use parsecs to measure distance D then the radius as in Fig. 6.3 equals 1 AU, so that for small angles Eq. (6.2) reads

$$\boxed{D(\text{pc}) = \frac{1}{\alpha \text{ (arcsec)}}}, \tag{6.3}$$

and thus, the distance to Barnard's Star can be expressed as $D = \frac{1}{0.547 \text{ (arcsec)}} = 1.83 \text{ pc} = 5.96 \text{ ly}$. It's that simple.[5]

Exercise
(a) The Dogstar Sirius A has a parallax of 379.2 mas. Compute its distance in light-years. *Ans.* 8.6 ly.
(b) The star Epsilon Eridani has a parallax of 311.37 mas. Compute its distance in light-years. *Ans.* 10.475 ly.

Exercise The black hole Cygnus X-1 exhibits a parallax angle of $\alpha = 0.4439$ mas. Compute the distance by:

(a) Using the average distance from the Sun to the Earth giving the distance in light-years;
(b) Using parsecs. *Ans.* 2.25×10^3 pc.

Exercise The star Mizar exhibits a parallax of 39.36 mas. Determine its distance. *Ans.* ~83 ly.

Galaxy Distance via Cepheids

Since we have just worked out the distance to nearby stars involving parallax, we need to now move further out to stars with no discernible parallax. This is achieved by observing a type of pulsating variable known as a Cepheid that has a regular fluctuation in brightness and are named after the prototype Delta Cephei

---

[5] More explicitly, $D = 1 \text{ AU}/\tan\alpha = 1 \text{ AU}/\alpha$. Now, 1 AU = 206,265 pc, and the original units of α were in radians. Converting this value of α to its equivalent value in arcsec: $\alpha \text{ (rad)} = \alpha \times 206{,}265 \text{ (arcsec)}$. Then

$$D = \frac{1 \text{ AU}}{\alpha \text{ (rad)}} = \frac{206{,}265 \text{ pc}}{\alpha \times 206{,}265 \text{ (arcsec)}} = \frac{1 \text{ pc}}{\alpha \text{ (arcsec)}},$$

which gives the distance D in parsecs when the parallax angle is measured in arcsec as per Eq. (6.3).

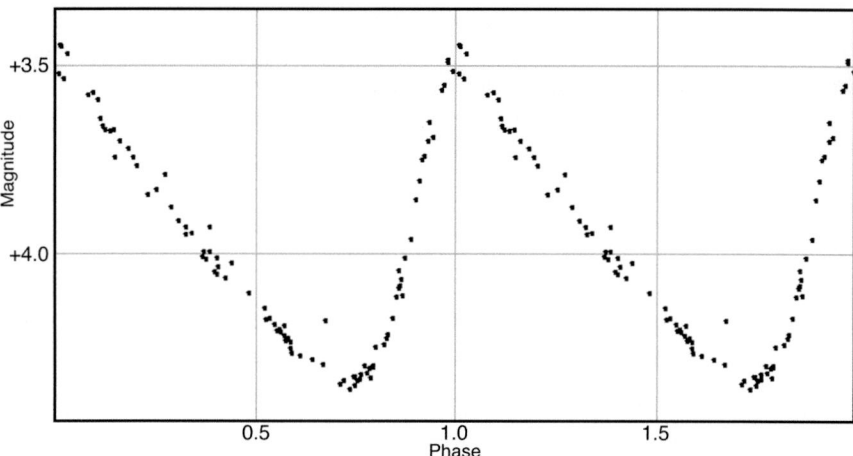

Fig. 6.4 The light curve for the variable star Delta Cephei whose visual light magnitudes vary over a period of 5.4 days which is indicated by 1 phase in the diagram. Note the gradual decline to the minimum followed by a rapid change to maximum brightness which is a typical feature of Cepheids. (Courtesy Jonathan Park)

(Population I[6]) whose visual light magnitude varies between 3.5 and 4.4 over a period of 5.366 days (1 phase). Such stars are 4 – 20 times the mass of our Sun. The closest cepheid variable is the North Star Polaris at 433 ly (Fig. 6.4).

But what do Cepheids have to do with distance? A remarkable discovery in 1912 by Henrietta Leavitt was published from data on 25 variables stars residing in the Small Magellanic Cloud.[7] Leavitt's observation was that there is a particular relationship between the brightness of the Cepheids and their periods (see Fig. 6.5). Leavitt made the very reasonable assumption that all the 25 variable stars were roughly at the same distance which meant that their periods must be related to their *intrinsic brightness*.[8] Essentially, Cepheids that are intrinsically brighter have longer periods and this can be expressed in a mathematically precise manner which will be given in the sequel (the *Leavitt Law*). This particular quality is why Cepheids are known as *standard candles*. A candle produces a specific light intensity and if it is further away it becomes dimmer and closer in it becomes brighter. Similarly, if two Cepheid variable stars have the same period and one appears brighter than the other,

---

[6] There are also Population II Cepheid variables which very old lower mass objects (~half the Sun's mass) with periods 1 – 50 days. But we will not make any distinction between the two populations nor the technical difficulties they present in distant measurements.

[7] Periods of 25 Variable Stars in The Small Magellanic Cloud, *Harvard College Observatory Circular* 173, 1912.

[8] The intrinsic brightness is the brightness of the star at its surface. It is due to the intrinsic qualities of the star itself. This is in contrast to the *apparent brightness* which is that observed from Earth.

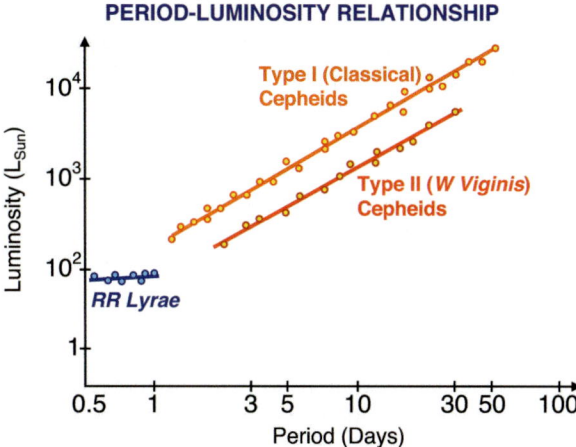

Fig. 6.5 The period-luminosity relation for Type I and Type II Cepheid variable stars. (From https://www.atnf.csiro.au/. © Copyright CSIRO Australia)

then we know it must be closer than the dimmer one. This feature allows for a determination of distance from the observer as will be seen below.

Apparent vs Absolute Magnitude

Before this information discovered by Leavitt is of any use to us (and Leavitt expressed this herself) we must somehow measure the distance to some Cepheid variables. This was done a year later by Ejnar Hertzsprung who measured the parallax to 13 Cepheid variables and hence their distance from Earth. So, now that we have the distance D to some Cepheid variables, we can measure their periods P and we can measure their (mean) *apparent magnitude m*. What is needed at this stage is some sort of proxy for a star's intrinsic brightness.[9] A star's apparent brightness is affected by its intrinsic brightness and its distance from us. Therefore, to compare the intrinsic brightness of any two stars, we place all stars by *fiat* at a distance of 10 parsecs (about 32.6 ly) and refer to this standardized brightness as their *absolute magnitude M*.

Next, let us recall from Eq. (1.7) just how the magnitude of one star is related to another, where

$$\frac{b_1}{b_2} = 100^{\frac{(m_2-m_1)}{5}},$$

where b_1, b_2 represent the apparent brightness of two stars with respective magnitudes m_1, m_2 and $m_1 < m_2$. Note the switch in indices just because we can.

---

[9]Technically, 'brightness' is measured in terms of luminosity, discussed in Chap. 3.

Furthermore, suppose that the two stars are one and the same, firstly at a distance D from the observer where it has apparent magnitude m, and then at a distance of 10 parsecs (which we assume is *less than* D^{10}) with an apparent magnitude M. Thus, the star at distance D (in pc from an observer) which has a magnitude m and let us say a brightness b_m, so that when the star is at the closer distance of 10 pc it has magnitude M and say a brightness, b_M. The reader is advised to draw for themselves a little picture here (the author does not wish to do everything for you). From the above equation, since $M < m$ and $b_m < b_M$

$$\frac{b_M}{b_m} = 100^{\frac{(m-M)}{5}}.$$

On the other hand, we have seen, the brightness of the light from the star will vary inversely proportionally to square of the distance from the observer which is to say

$$\frac{b_M}{b_m} = \frac{D^2}{10^2}.$$

We now equate the above two equations which yield

$$100^{\frac{(m-M)}{5}} = \frac{D^2}{10^2},$$

and taking the log base 10 of both sides

$$(m-M)\frac{2}{5} = 2\ \log\left(\frac{D}{10}\right).$$

Therefore, the preceding equation now becomes,

$$m - M = 5\ \log_{10}\left(\frac{D}{10}\right).$$

This formulation is absolutely fine, but let us go one step further and evaluate the *log* of the quotient,

$$m - M = 5(\log_{10} D - 1),$$

or as it is customarily written

---

[10] This assumption has no significance and only made to maintain the order of the respective variables.

Galaxy Distance via Cepheids

$$\boxed{m - M = 5\log_{10} D - 5} \tag{6.4}$$

where D is the distance to the star in parsecs.[11]

The distance D is determined by parallax and the apparent brightness m is observed by telescope. Now we can easily determine the absolute magnitude M (our proxy for intrinsic brightness) which is the essential ingredient in the period-luminosity relationship determined by Leavitt when the distance is known. But the real virtue of this formula is when the distance of the star is not known and we happen to know the absolute magnitude, M.

Period-Luminosity Relation

There are two basic classes of Cepheid variable stars, Type I (classical) which are usually found in the disk of a galaxy and are young (10 – 300 million years old), and more luminous than Type II Cepheids which are older (>10 billion years old) and found in the halo and bulge of galaxies as well as in globular clusters. The Type I Cepheids are the ones usually used for distance calculations although both types exhibit a period-luminosity relation (Fig. 6.5).

Often the absolute magnitude is used as a proxy for the luminosity since we know they are related by Eq. (3.11): $M = -2.5 \log\left(\frac{L}{L_\odot}\right) + M_\odot$. Data derived from Cepheids from the Milky Way and LMC gives the best fit straight line relating the absolute magnitude M to the period P as

$$\boxed{M = -2.78\log_{10} P - 1.22}. \tag{6.5}$$

This formula is derived from the data given in Fig. 6.6 via linear regression discussed in Chap. 1 and is known as the *period-luminosity (P-L) relation*. The M in the data represents the absolute magnitude . It must be said that various other sources give slight variations on this formula so the above is not the last word on the subject. There are many such PL-relations derived from other data sets. Fortunately, most agree within very reasonable limits.

Note that for $P = 10$ days then $\log_{10} 10 = 1$ and the M value given by the regression line is -4.00.

For example, if we take an actual Cepheid from the Large Magellanic Cloud, HV 2549[12], then its period is $P = 16.22$ days. Therefore, $\log_{10} P = 1.21$ and by Eq. (6.5)

---

[11] The formula also takes into account obscuring dust and gas.
[12] All data for HV2549 taken from P. Karczmarek et al., Large Magellanic Cloud Cepheids in the ASAS data, *Acta Astrometrica* 61 (2011), 303–318.

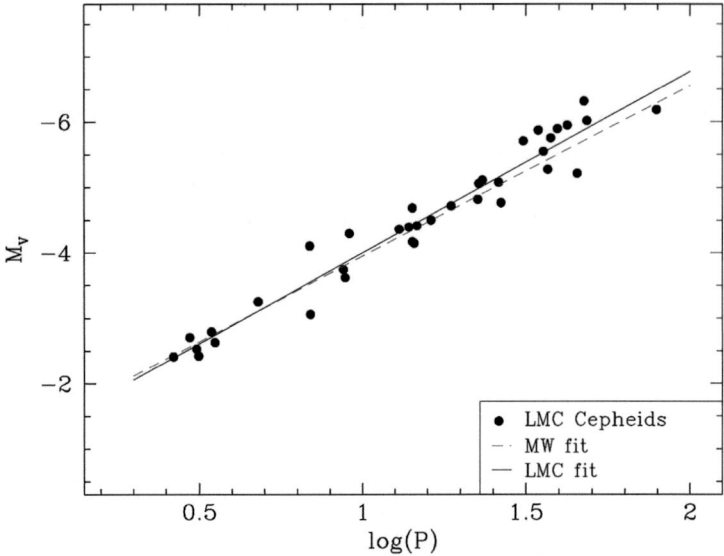

Fig. 6.6 Here M_V is the absolute visual magnitude depicting the period-luminosity relation. The best regression line fit of the data is given $M = -2.78 \log_{10} P - 1.22$. (Courtesy Storm et al., Calibrating the Cepheid Period-Luminosity relation from the infrared surface brightness technique, II, *A&A*, 534 (2011), 11 pp.)

$$M = -2.78 \times 1.21 - 1.22$$
$$= -4.58,$$

which is the star's absolute magnitude and consistent with Fig. 6.6.[13]

Distance Modulus

Once the value of M has been determined, then we can go back to Eq. (6.4) and use that formula to determine the distance D (in parsecs) since all the other parameters are known at this stage. Solving Eq. (6.4) for the parameter D, we obtain

$$D = 10^{\frac{m-M+5}{5}}. \quad (6.6)$$

Using our Cepheid variable star HV2549, with $M = -4.58$ and a mean visual magnitude of $m = 13.96$ we can solve for the distance

---

[13] Another P-L relation is: $M = -2.43 \log_{10} P - 1.62$. When applied to the example in the text of HV 2549, it yields an absolute magnitude $M = -4.56$. This is very close to the value in the text of $M = -4.58$. **Exercise.** Compute the distance to the LMC using this new value for M. Ans. ~163,000 ly.

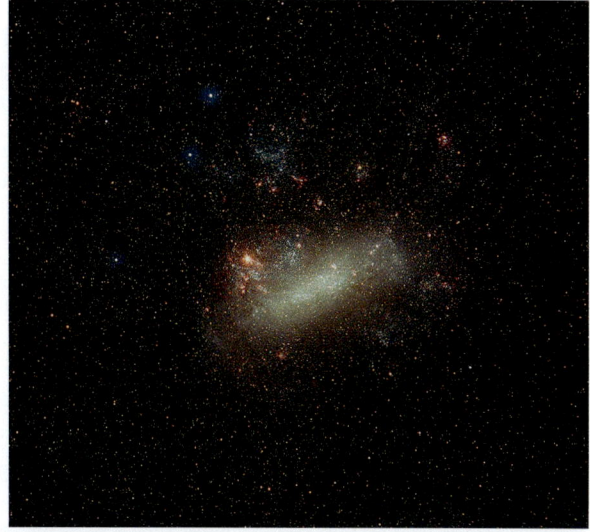

Fig. 6.7 A galaxy readily visible with the naked eye from the Southern Hemisphere is one of our nearest neighbors, the Large Magellanic Cloud at only 163,000 light years distant. (Courtesy Eckhard Slawik, ESA/Hubble. e. slawik@gmx.net)

$$D = 10^{\frac{m-M+5}{5}} = 10^{\frac{13.96+4.58+5}{5}} = 10^{4.71}$$

$$= 51{,}286 \text{ parsecs} \sim 167{,}000 \text{ ly}.$$

This value is in close agreement with the value quoted in the literature[14]. Instead of taking 1 or 2 variable stars one needs to consider numerous stars and take their average distance for scientific rigor. The all-important quantity in this type of calculation is: $m - M$, which is called the *distance modulus* and for the visual band is written as $\mu = (m - M)_0$, where the subscript naught indicates that this figure has been corrected slightly for interstellar absorption which becomes more of a factor at larger distances.

In our case, the distance modulus for the LMC $= 18.54$. That is really the determining factor in the distance calculation and many astronomical papers only quote that value since the rest is just a routine calculation. Indeed, many research articles in the literature take $(m - M)_0 = 18.5$ for the LMC as their starting point (Fig. 6.7).

Exercise Another Cepheid variable star in the LMC is HV2827 which has a visual magnitude of $m = 12.28$ and a period of 78.70 days.

(a) Calculate its absolute magnitude M. *Ans.* $M = -6.26$.
(b) Determine the distance to the LMC via this variable star. Ans. ~50,000 pc $= 163{,}000$ ly.

---

[14] A value of ~163,000 ly is a reasonable average value and corresponds to $(m - M)_0 = 18.5$.

In 1923, Edwin Hubble located his first Cepheid variable star in the Andromeda Galaxy which has famously become known as M31-V1.[15] He measured its period and its mean apparent magnitude to determine the distance to Andromeda which had to place it outside the Milky Way contrary to what was thought at the time.

Using a period of V1 of $P = 31.39$ days, let us try the P-L relation from footnote 13: $M = -2.43\log_{10} P - 1.62$, so that the absolute magnitude is given by

$$M = -2.43 \times \log_{10}(31.39) - 1.62 = -5.26.$$

V1's mean apparent magnitude is listed at $m = 19.2$ so that the distance modulus is

$$(m - M)_0 = 24.46$$

which is the same (or very close to) the value obtained by many other researchers.[16] This also works out to ~2.54 million light-years. See Fig. 6.8.

Fig. 6.8 The Cepheid variable star V1 in the Andromeda Galaxy M31 taken by the Hubble Space Telescope. (Courtesy NASA, ESA, Hubble Heritage Team)

---

[15]"V1 is the most important star in the history of cosmology," according to astronomer Dave Soderblom of the Space Telescope Science Institute (STScI) on a NASA website.

[16]The website: ned.ipac.caltech.edu/ lists 407 modern determinations of the distance to M31.

There are other regularly variable stars such as RR Lyrae stars commonly found in globular clusters that have also been used for distance calculations.

Type 1a Supernovae

There will be a point when a galaxy is simply too distant in order to work on individual stars and other means must be called upon. One very widespread and useful method discussed in the next chapter is based on the velocity of recession of the galaxy. But let us mention another based on a *Type 1a supernova* explosion which are also known as standard candles. The classic progenitor is a white dwarf approaching ~1.44 $M_\odot$ (although there can be some variation from this value known as the *Chandrasekhar limit* discussed in Chap. 8) and its peak visual (and blue band) absolute magnitude is typically $M \approx -19.3$.[17] The light intensity is so great that it can be seen across billions of light-years and has provided the data that the Universe's expansion is accelerating (Fig. 6.9).

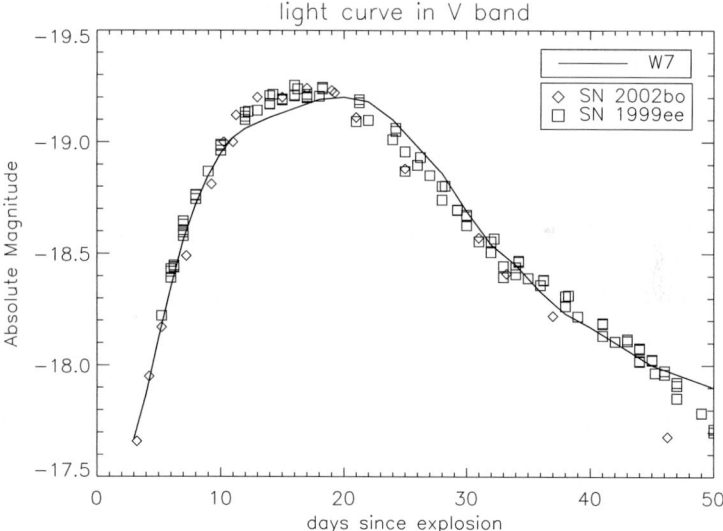

Fig. 6.9 The characteristic curve of a Type 1a supernova with a rapid rise to maximum brightness over a period of about 20 days exhibited by SN1999ee and SN2002bo followed by a rapid decline and then a gradual diminution of brightness. The solid curve is from a computer model. (Courtesy Dennis Jack et al., Theoretical light curves of type Ia supernovae, *A&A*, 528 (2011), A141)

---

[17] Some sources take the value of $M = -19.5$, and there can be slight variations due to progenitor star differences.

For example, the Type 1a supernova SN 2011fe was detected in the Pinwheel Galaxy M101 and reached a peak apparent magnitude of $m = 9.9$. Then its distance modulus is

$$(m - M)_0 = 9.9 + 19.3 = 29.2.$$

Of course, as with any such determination, we really need more data than a single supernova. But the above agrees well with values in the literature and gives a distance to M101 of 22.6 Mly.

Regarding the acceleration of the expansion of the Universe, ever since Edwin Hubble's confirmation that the Universe is expanding, it was generally believed that this expansion would gradually slow down from the effects of gravity due to the mass density of the Universe. This was put to the test by two different groups of astronomers in 1998 who found that at given redshifts (which determines velocity and distance) the Type 1a supernovae were dimmer than their redshift distance indicated, and hence further away than their redshift indicated, and consequently receding faster than expected. The conclusion was drawn that the expansion of the Universe is accelerating, a topic that will be discussed further in Chap. 10 where we explore what could be driving the cosmic acceleration. For this discovery, the group leaders, Adam Riess, Saul Perlmutter, and Brian Schmidt won the 2011 Nobel Prize in Physics for their remarkable discovery.

Tip of the Red Giant Branch

Another rather exotic sounding standard candle is known as the *tip of the red giant branch* (TRGB). These are aging red giant stars that have expanded and cooled to about 3000 K reaching a critical tipping point of the Hertzsprung-Russell diagram of temperature vs luminosity. Such stars have a standard infrared (I band) absolute magnitude of $M_i = -4.05$[18] and so can be used to measure distances in the same manner as Type 1a supernovae. See the distance calculation for the galaxy M87 in Chap. 9 using this method.

---

[18] Freedman et al., Calibration of the Tip of the Red Giant Branch, *ApJ* 891 (2020), 30pp.

Tully-Fisher Relation

It was observed that there is a relation between the intrinsic luminosity of a spiral galaxy and the maximum rotational velocity of its gas or stars in the galaxy's disk. The important relation (due to R. B. Tully and J. R. Fisher[19]) has the power law form

$$L \propto V_{max}^{\alpha}$$

where $\alpha \approx 4$ depends on the type of spiral galaxy. This then takes the form

$$\boxed{\log L = \alpha \log V_{max} + \beta} \tag{6.7}$$

so that the log-log graph of this relationship becomes a straight line with the slope given by the power α and the value of β is the y-intercept. In this context, what is used in practice for V_{max} is the flatlined rotational velocity, V, which is a stable proxy measure across different galaxies and not influenced by any local mass variations (Fig. 6.10).

As well as the luminosity, the galaxy's baryonic[20] mass can also be determined. In this formulation the relation is known as the *Baryonic Tully-Fisher relation* (BTF)

$$\boxed{M = \alpha \log_{10} V_{max} + \beta}. \tag{6.8}$$

In this instance, the power α is again close to four which implies that the baryonic mass of the galaxy is roughly proportional to the fourth power of the maximum rotational velocity (see Fig. 6.11). Note that in this instance the variable 'M' stands for mass, not absolute magnitude.

Instead of the (intrinsic) luminosity L a proxy of *absolute magnitude* M[21] is used in practice where the relation between L and M is

$$\boxed{M = -2.5 \log\left(\frac{L}{L_0}\right) + M_0} \tag{6.9}$$

where L_0 and M_0 are the luminosity and absolute magnitude of a well-known standard reference galaxy. This is just the same stellar relation as was derived for

---

[19] A new method of determining distances to galaxies, A&A. 54 (1977), 661–673. Their work used the Doppler shift of the 21-cm line of neutral hydrogen measured via radio telescope to determine the velocity of gas rotation instead of stellar redshift. See Chap. 7.
[20] This is essentially the normal mass of stars and gas. The mass of electrons is ignored in this terminology but is of little consequence.
[21] As with stars, the absolute magnitude is its apparent magnitude if the galaxy were located at a distance of 10 parsecs.

Fig. 6.10 Graph of the best fit rotational velocities at various distances from the center of the giant spiral galaxy 2MFGC08638 showing the flatlined velocity that is used in the Tully-Fisher relation. Vertical lines indicate uncertainty in measurements and empty circles are inferred velocities from the data. (Courtesy Enrico M. Di Teodoro)

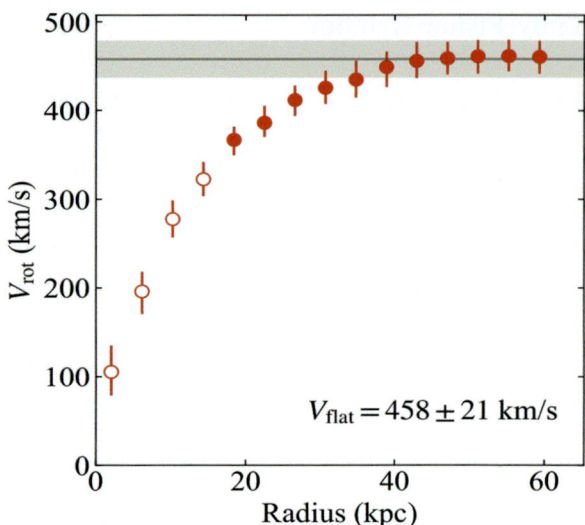

Fig. 6.11 The Baryonic Tully-Fisher relation log $M = \alpha \log V_{flat} + \beta$ relating the (blue band) baryonic mass M_b of spiral galaxies to their flatlined rotational velocity V_f. The dark blue circles represent star-dominated galaxies; light blue circles are galaxies having more mass in gas than in stars. The slope is $\alpha = 4$. (Courtesy Stacy McGaugh)

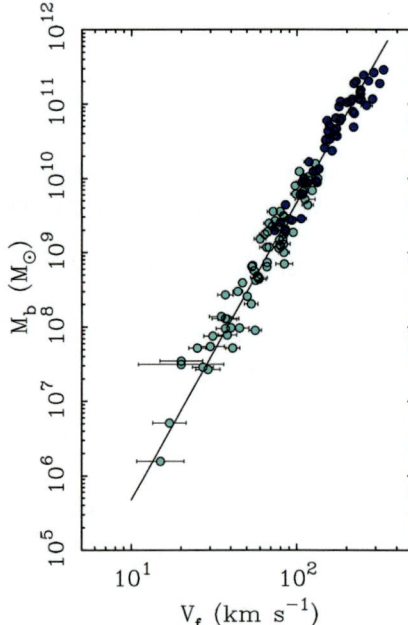

Eq. (3.11). With L now determined by Eq. (6.7) we obtain the relation in terms of absolute magnitude

Tully-Fisher Relation

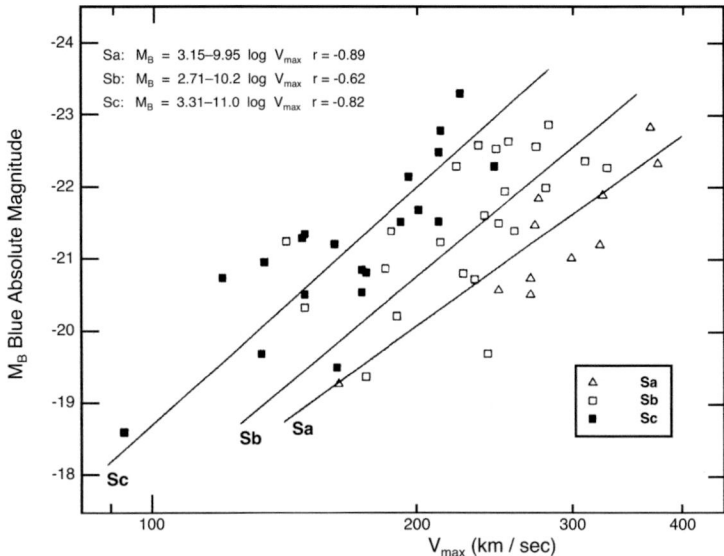

Fig. 6.12 Graphs of the Tully-Fisher relation for various spiral galaxy types (Sa, Sb, Sc). Each type gives a slightly different linear relation. (Courtesy *The Astrophysical Journal*, from V. Rubin et al., Rotation velocities of 16 Sa galaxies and a comparison of Sa, Sb, and Sc rotation properties, *Ap. J.*, 289 (1985), 81–104)

$$\boxed{M = -2.5\alpha \, \log \, V_{max} + C}. \tag{6.10}$$

where M is the absolute magnitude (!) of the galaxy and C is a constant arising from β, L_0 and M_0 and has no significance. *All values are with respect to a particular band* such as the B-band, I-band, or V-band.

In the data in Fig. 6.12 given in the blue band, it is clear that the constants α and C depend on the type of spiral galaxy in question. From the graphs one observes that the spiral galaxies with higher rotational velocities have higher absolute magnitudes. From the data for Sa, Sb, and Sc spiral galaxies, one can divide the respective coefficients of -9.95, -10.2, and -11.0 by -2.5 in Eq. (6.10) to tease out the corresponding coefficient α, all of which are close to 4 also as seen in Fig. 6.11.

The Tully-Fisher relation is an important distance determinator for a galaxy and indeed their 1977 paper was about determining distances. Once the luminosity and hence absolute magnitude M has been determined for a galaxy as above (in a specific band), then the apparent magnitude m is measured directly (in the same band). The distance modulus formula (Eq. (6.4) is then used to compute the distance to the galaxy in parsecs, namely

$$\boxed{m - M = 5\log_{10} D - 5}. \tag{6.11}$$

Again, one then merely has to solve for the unknown quantity D which is the galactic distance.

One of the most common distance determinators is via the redshift discussed in the next chapter.

Galaxy Diameter

Once the distance D to a galaxy has been determined by whatever means, it is possible to determine its diameter d (in the plane of the sky) using the formula gleaned from Fig. 6.13 where θ is in radians,

$$\frac{d/2}{D} = \tan\frac{\theta}{2}, \text{ or } d = 2D\tan\frac{\theta}{2}.$$

Since we know that for small angles θ,

$$\tan\frac{\theta}{2} = \frac{\theta}{2},$$

it follows that

$$d = 2D \times \frac{\theta}{2} = D\theta.$$

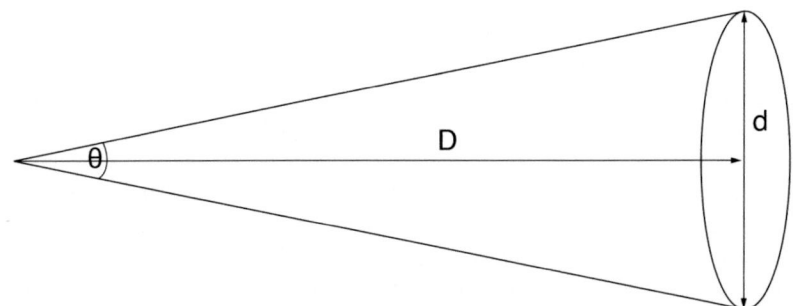

Fig. 6.13 Knowing the distance D to a galaxy and measuring its angular diameter θ, then the galaxy's diameter d can be calculated from $d = D\,\theta$. (Courtesy Katy Metcalf)

Fig. 6.14 For very small subtended angles the value of the diameter *d* is approximately the same as the length *s* of the arc of a circular sector. (Courtesy Katy Metcalf)

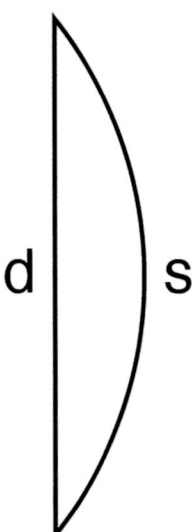

We can view this in another way and note that regarding the Chap. 1 formula $s = r\theta$, (Fig. 1.2) for the length s of the arc of a circular sector of radius r having angle θ, it will be the case that for *small values*[22] of θ that $s \approx d$ as in Fig. 6.14.

Since in our notation $r = D$, we obtain as above

$$\boxed{d = D\theta}. \tag{6.12}$$

The angle θ is the telescopically measured angular diameter (initially in fractions of a degree, then converted to radians) of the galaxy. If we are just dealing with the radius of a celestial ring, then $r = d/2$ and

$$r = D \cdot \theta / 2$$

where the angle θ is again as in Fig. 6.13 and $\frac{\theta}{2} = \alpha$ is the *angle subtended by the radius*.

For example, the beautiful Needle Galaxy NGC 4565 (see Fig. 6.15) spans an angular size of 15.9 arcmin $=0.004625$ rad. Its distance has been measured to be 55 Mly, so that its diameter is

$$d = (55 \times 10^6) \times (4.625 \times 10^{-3}) \sim 254{,}000 \text{ ly,}$$

which is roughly two-and-a-half times the diameter of the Milky Way ($\sim$100,000 ly).

---

[22] For angles up to about 4 or 5 degrees.

Fig. 6.15 The Needle Galaxy at a distance of 55 million ly with a diameter of approximately 254,000 ly. (Courtesy ESO)

Exercise
(a) Compute the diameter of the galaxy M74 that has an angular diameter of 10.2 arcmin and is at a distance of 32 million light-years. *Ans.* 95,000 ly.

Exercise The supernova 1987A explosion in the LMC illuminated a ring of previous ejected material via UV radiation after a period of 240 days, the radiation travel time (at the speed of light).

(a) Assuming the progenitor star and the circular ring lie in the plane of the sky, compute the radius of the ring in km. *Ans.* 6.22×10^{12} km.
(b) The ring has an angular radius as viewed from Earth of 0.808 arc. Converting this to radians and using part (a), compute the distance to SN1987A in the LMC in light-years. *Ans.* 168,000 ly. This value compares favorably with our former determination of 167,000 ly using the Cepheid variable HV 2549, and the average quoted distance of 163,000 ly.

Chapter 7
Hubble's Law of the Universe

Doppler Effect/Redshift

Lying at the core of many astronomical calculations is a natural phenomenon first described by the Austrian scientist Christian Doppler in 1842. In Astronomy the *Doppler Shift* is the change in wavelength/frequency of electromagnetic radiation due to the relative motion of the source and observer. This is a common occurrence heard when the siren of an emergency vehicle like an ambulance sounds higher in pitch as it approaches and lower as it moves away. Since the Universe is expanding, our position on Earth will be considered stationary with respect to this recession although there is the rotation of the Earth at 0.46 km/s and the mean orbital velocity of the Earth around the Sun at 29.78 km/s to consider, but these will be set aside as they turn out to be insignificant when dealing with the celestial velocities in this book. As will be seen in Chap. 8, gravity also affects the wavelength of radiation leaving a massive body, but this too is a very minor effect and will be similarly ignored.

As with a receding ambulance whose siren sound becomes lower in pitch (frequency), that is, longer in wavelength, the same is the case with light from receding galaxies. Their visible light will become longer in wavelength, which means it will be shifted towards the red end of the spectrum known as a *redshift*. In some cases where the light source is approaching the observer, the light will be shifted towards the blue end of the spectrum, a *blueshift*. Both phenomena occur observing the rotation of a spiral galaxy when side of the galaxy is receding from us while the other side is approaching. Moreover, in most of the cases that we will consider, the amount that the light has been shifted is directly proportional to the velocity of recession and the constant of proportionality is the velocity of light itself, c. So, what could be nicer.

But this begs the question: How do we measure the amount that light has been red or blue shifted? As it turns out, when observing the spectrum from moving celestial bodies such as stars or galaxies, the normal rainbow-colored spectrum is interrupted

Fig. 7.1 The continuous, absorption, and emission spectra obtained by viewing the light source under differing conditions. (Courtesy Katy Metcalf)

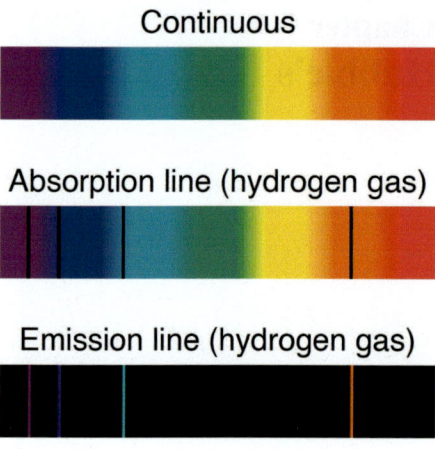

by certain black lines. These black spectral lines were used by English astronomer William Huggins in the 1860s to demonstrate that the stars are receding from or moving towards the Earth. Thus, we need to look at the sources of these spectral lines and their significance.

The spectral lines occur from two basic scenarios. In the first instance, on its passage to Earth, if the light from a relatively hot star passes through a thin cooler gas, then the atoms and molecules in the gas absorb photons of light at specific wavelengths. This occurs when the light's energy matches the difference between two energy levels in the atoms or molecules of the gas which results in a dark line at the wavelength of the absorbed photons. This produces an *absorption spectrum*. An *emission spectrum* is produced in objects such as nebulae or in the atmosphere of certain stars when atoms or molecules are excited by say collisions or radiation and upon returning to a lower energy state emit light at specific wavelengths which results in bright lines (emission lines) in the continuous spectrum. See Fig. 7.1. Either absorption or emission lines can be used for the purposes of the Doppler shift and occur in the same positions on the light spectrum.

The first order of business is to quantitatively measure the amount the spectrum has been shifted from its rest state appearance in the laboratory. This is achieved by measuring the shift in any one of the absorption lines in a celestial object's spectrum and compare it to its original position in a lab spectrum. For example, the absorption lines in the visible spectrum produced by hydrogen gas are at 410 nm 434 nm, 486 nm, and 656 nm,[1] each one corresponding to an electron jumping to a higher energy state in the absorbing hydrogen medium. The amount of the spectral redshift denoted by z is given by

---

[1] Other absorption/emission lines occur in the non-visible part of the spectrum for hydrogen.

Doppler Effect/Redshift

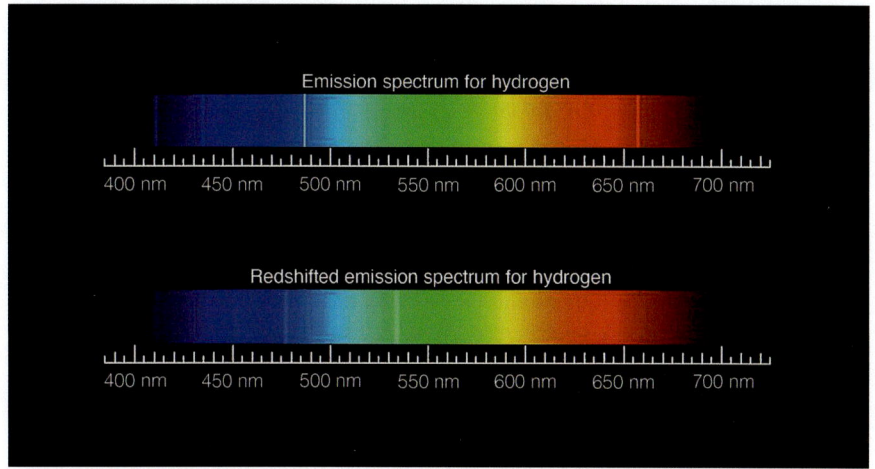

Fig. 7.2 Top: The laboratory emission spectrum for hydrogen at 410 nm, 434 nm, 486 nm and 656 nm. Bottom: The redshift of a receding galaxy compared to the laboratory reference. (Courtesy NASA/JPL-Caltech)

$$z = \frac{\Delta \lambda}{\lambda_{sta}} = \frac{\lambda_{obs} - \lambda_{sta}}{\lambda_{sta}}, \quad (7.1)$$

where λ_{obs} is the observed Doppler-shifted wavelength of an absorption line, and λ_{sta} is the unshifted wavelength of the corresponding absorption line from a stationary emitter. The value of z has no units attached and merely serves to quantify the shift in spectral lines of a moving source compared to a stationary one. It has been found that the vast majority of galaxies in the Universe are moving away from us, although a nearby neighbor, Andromeda, and the Milky Way are actually approaching one another resulting in a blue shift in Andromeda's light. This will result in a cosmic embrace in about 4–5 billion years from now.

The redshift can also be expressed in terms of frequency since $\lambda = c/f$, and thus we can write Eq. (7.1) as

$$z = \frac{f_{sta} - f_{obs}}{f_{obs}}.$$

In order to calculate the redshift for an object in space it just requires that the change in wavelength be measured between any two corresponding absorption lines. For example, let us take the reference absorption line of hydrogen at the (third) position of $\lambda_{sta} = 486$ nm, whereas and the same line again for a receding (Fig. 7.2) galaxy is at $\lambda_{obs} = 534$ nm so that

$$z = \frac{534 - 486}{486} = 0.0988$$

The units can be left out as they will cancel anyway. Blue shifted objects like Andromeda will give a negative value of z in view of Eq. (7.1).

For small values of redshift, that is say for, $z < 0.15$, the velocity of recession of the object omitting the light is given by

$$\boxed{v_r = cz}. \tag{7.2}$$

This has the interpretation that the redshift represents the fraction of the speed of light at which the object is receding from us, namely v_r. So in the above example, the galaxy is receding at about one-tenth the velocity of light. While considering values of $z < 0.15$ might seem overly restrictive, a great many galaxies fall into this category.

As an example, Needle Galaxy NGC 4565 has a redshift $z = 0.004103$ so that

$$v_r = 300{,}000 \times 0.0041 = 1{,}230 \text{ km/s}.$$

Considering that a rifle bullet travels at roughly 1 km/s, both galaxies are receding extremely rapidly.

The commonly used formula for obtaining the Doppler redshift velocity v_r as given by Special Relativity is slightly more complex,

$$z = \sqrt{\frac{1 + \frac{v_r}{c}}{1 - \frac{v_r}{c}}} - 1, \quad \text{or } (1+z)^2 = \frac{c + v_r}{c - v_r}.$$

Solving this expression for v_r/c we obtain

$$\frac{v_r}{c} = \frac{(1+z)^2 - 1}{(1+z)^2 + 1} \tag{7.3}$$

that is often used for recessional velocity determinations. Indeed, this formulation agrees with $v_r = cz$ at low values of z as the reader can verify, since if we take say $z = 0.10$ and put it into the preceding equation then we obtain $v_r/c = 0.10 = z$ (rounded off). Observe that in the formulation Eq. (7.3), as z becomes arbitrarily large, the value of v_r/c is approaching 1, that is, v_r is flatlining at the value c.

However, both formulations of Eq. (7.2) and Eq. (7.3) are *only suitable for low redshift values*. Note that the formulation in Eq. (7.3) is always less than 1 for any redshift value of z since the denominator will always be greater than the numerator. Therefore, in this formulation recessional velocities will always be less than c which is not the case at all as indicated by Fig. 7.3.

Therefore, for redshifts greater than about $z = 0.3$ we need to take General Relativity into account (and not Special Relativity). This is beyond the scope of

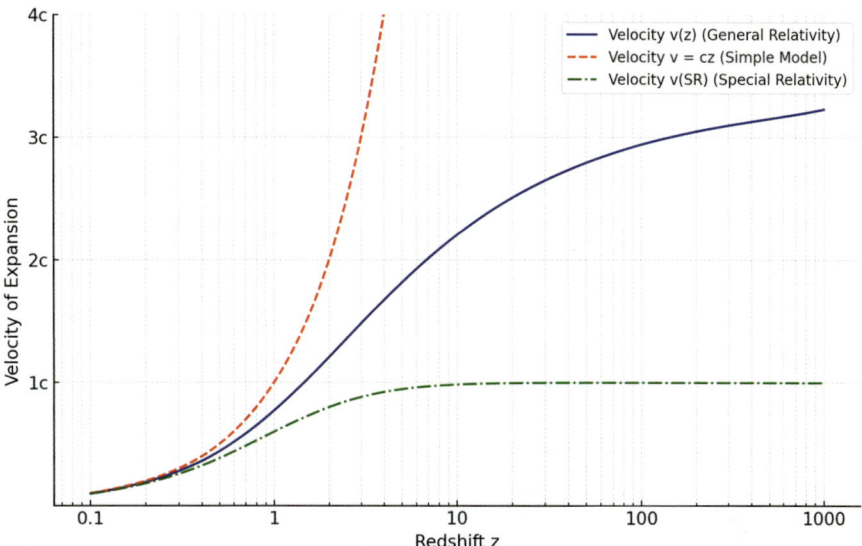

Fig. 7.3 Graph indicating recessional velocity for a standard model of the Universe incorporating General Relativity vs. the values given by the formulas $v_r = cz$ and the formula of Eq. (7.3) from Special Relativity. Adapted from Tamara M. Davis and Charles H. Lineweaver, Expanding Confusion: Common Misconceptions of Cosmological Horizons and the Superluminal Expansion of the Universe, *Publ. Astron. Soc. Australia*, 21 (1), 2004, 97–109. (Image created with ChatGPT4 by the author)

this book but there are online calculators for determining velocity as a function of redshift such as in Ned Wright's: https://astro.ucla.edu/~wright/CosmoCalc.html

It is evident from the graph (Fig. 7.3) that the recessional velocity of high redshift distant galaxies can be greater than the speed of light, even at the time the light was emitted. However, we are still able to observe these objects because the emitted photons eventually entered regions of space where their progress towards us outpaced the Universe's expansion. Such is the case with the galaxy JADES-GS-z14-0 with a recorded redshift of 14.32 discovered by the James Webb Space Telescope in 2024 and receding from us at more than twice the speed of light. Indeed, all galaxies with a redshift of $z > 1.46$ (in the current standard model of the Universe) are receding faster than the speed of light.

In fact, it is the space between the source and the observer that is expanding faster than the speed of light, which does not contradict the Theory of Relativity, as discussed in Chap. 8. This is because space itself can expand faster than light. Consequently, recessional velocity becomes a less useful concept for high redshift values, and the notion of distance requires further clarification, which we will discuss in Chap. 10.

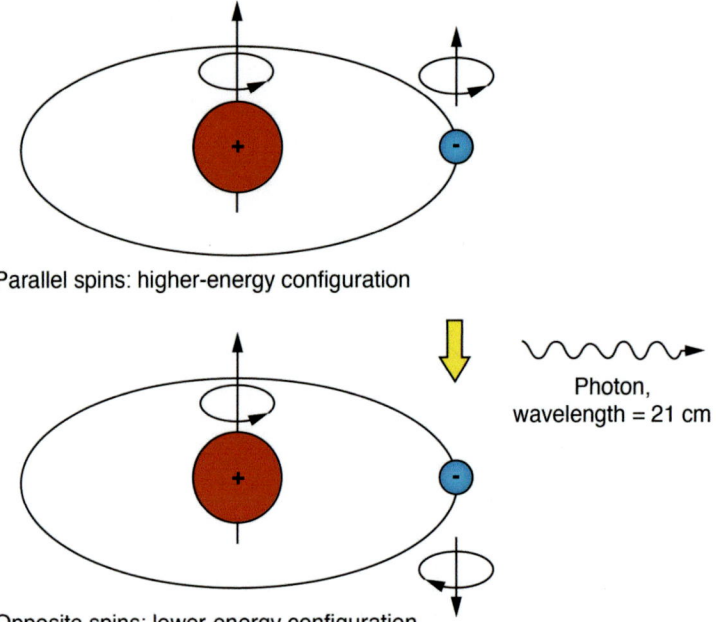

Fig. 7.4 When an electron of neutral atomic hydrogen spontaneously flips its orientation from being parallel with the proton to an anti-parallel orientation, a photon is given off having wavelength 21 cm (1420 MHz). (Courtesy Katy Metcalf)

21-cm Neutral Hydrogen

Not only visible light is redshifted by receding celestial objects, but it is also the case that one can use other electromagnetic waves that exhibit a Doppler shift. This is a very effective tool for radio astronomers. Very occasionally the electron of a hydrogen atom will reverse its spin and in so doing it releases a photon at a wavelength of 21 cm (1420 MHz) due to the small energy difference between the two states (Fig. 7.4). Although this is a rare event that can be brought on by collisions with other particles/atoms, or even spontaneously,[2] the shear large number of hydrogen atoms in a gas cloud permits their detection. The formulas are the same as for visible light Doppler shifts and this allows for the study of the velocity of gas clouds within galaxies for low z values.

However, now we wish to turn our attention to distance, and in order to say what we mean by 'distance' let us declare that it represents the *light travel time distance* from the time the light was emitted to when it was observed on Earth. Hence, if the distance of an astronomical body is given at say 100 million light-years, that means its light has been traveling for 100 million years to reach us even though the

---

[2] After ~10 million years.

Universe has expanded in the meantime and the object is now much further away. We know how to measure distance when there are Cepheids to be seen which is the first rung on our distance ladder, but when a galaxy is too far away to see Cepheids or we are looking at a quasar, we need to employ other tools.

Hubble Parameter/Constant

The American astronomer Edwin Hubble made observations of Cepheid variables in the early 1920s and working with redshifted velocities he plotted those velocities against their distances. The result was that the data all aligned along a straight line meaning that the relationship between distance and recessional velocity was a linear one, that is

$$\boxed{v_r = H_0 D} \tag{7.4}$$

where D is the distance (in Mpc) that is determined from our present-day image of the object and H_0 is just the constant multiple of D that determines the velocity, v_r (in km/s). This recessional velocity actually represents the expansion of the Universe and is in a radial direction for all observers. The term H_0 indicates the Hubble Parameter for the present era, which we take as constant, and its significance will be discussed in the next chapter.

This simple linear equation (Eq. 7.4) is known as *Hubble's Law* and is one of the most fundamental cosmological relations of the Universe.

The data from numerous galaxies is given in Fig. 7.5 and gives a value of 73 km/s/Mpc.[3] That is to say, with every additional million parsecs of distance, the Universe is expanding an additional 73 km/s.

If we happen to know the recessional velocity, say from the redshift or other means, then we have the formula for the distance (in Mpc),

$$\boxed{D = \frac{v_r}{H_0} = \frac{cz}{H_0}}, \tag{7.5}$$

again, with the proviso that v_r represents the redshift velocity for small redshift values.

The computed value of H_0 over the years has fluctuated considerably and that is one of the reasons for differing distance estimates to celestial bodies. The formula is

---

[3] There has been some variability in the precise value of the Hubble Constant ranging to 67 to 74. We will use the value of 73 from: A.G. Riess, A Comprehensive Measurement of the Local Value of the Hubble Constant with 1 km s^{-1} Mpc^{-1} Uncertainty from the Hubble Space Telescope and the SH0ES Team, *Astrophys. J. Lett.*, 934 L7 (2022), 55 pp. The value determined was $H_0 = 73.04 \pm 1.04$ km/s/Mpc.

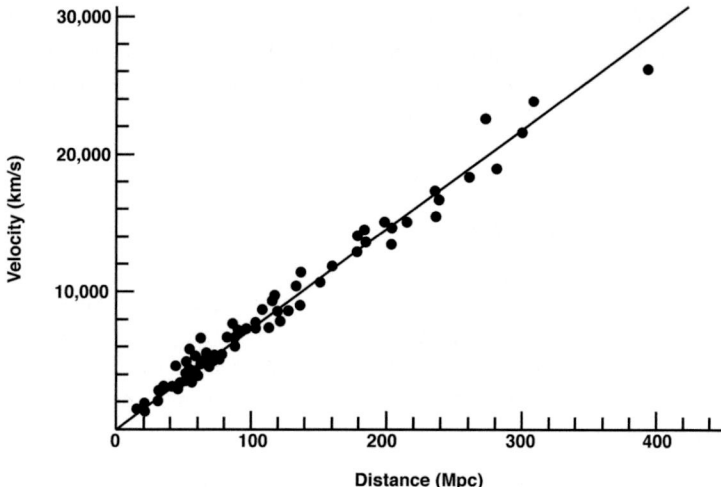

Fig. 7.5 The Hubble Constant derived from data from numerous galaxies is given by the slope of the line which has a value $H_0 = 73$ km/sec/Mpc. At $v_r = 20{,}000$ km/s, the distance is ~ 274 Mpc. (Credit: ASU and Katy Metcalf)

valid for a wide range of recessional velocities but becomes less accurate at very large distances and very high velocities. It should also be noted that the distance given by the distance modulus in Eq. (6.4) is in *parsecs* so if we combine that equation with Eq. (7.5), we must be very careful.

As an example, let us consider again the Needle Galaxy NGC 4565 which as we saw above has a redshift recessional velocity of $v_r = 1230$ km/s. Then the distance to the Needle Galaxy is

$$D = \frac{1230}{73} = 16.85 \text{ Mpc} = 55 \text{ Mly}.$$

Exercise The redshift for the galaxy NCG 6861 (Fig. 7.6) is $z = 0.009437$. Determine its distance. *Ans.* 126 Mly.

Exercise
(a) Express the distance modulus $m - M$ in terms of redshift. *Ans.* $m - M = 5 \log_{10}\left(\frac{cz}{H_0}\right) + 25$.
(b) Compute the distance modulus of a galaxy with redshift $z = 0.2$, taking $H_0 = (73 \text{ km/s})/\text{Mpc}$. *Ans.* $m - M = 39.57$ as per the graph in Fig. 7.7.

Figure 7.7 is real data of the distance modulus $\mu(z) = m - M$ vs redshift. The formula in part a) is best suited to low redshift values and a more sophisticated version of the formula best suits the higher redshift values.

Fig. 7.6 The lenticular galaxy NGC 6861 in the constellation Telescopium. It is thought that this galaxy will merge at some stage with nearby galaxy NGC 6868. (Courtesy ESA/Hubble & NASA Acknowledgement: J. Barrington)

(c) What is the distance to the galaxy in part (b)? *Ans.* ~821 Mpc.

Lookback Time

As mentioned above, in our reference to celestial distances, we are considering the light travel time distance, and as a proxy we can just refer to the *lookback time*, that is, the time the light took to reach us from the object in question. Then

$$\text{distance} = c \times \text{lookback time}.$$

Of course, slightly varying distances to celestial bodies have been given in the literature over the years by using a different value for H_0 as well as different observational techniques to determine distance such as Type 1a supernovae, tip of the red giant branch, etc.

We know from the Hubble Law that galactic distance increases with recessional velocity and that the latter increases with redshift, which means that the lookback time is also a function of redshift as illustrated in Fig. 7.8. Here we can go out to high redshift values but the relationship is no longer linear.

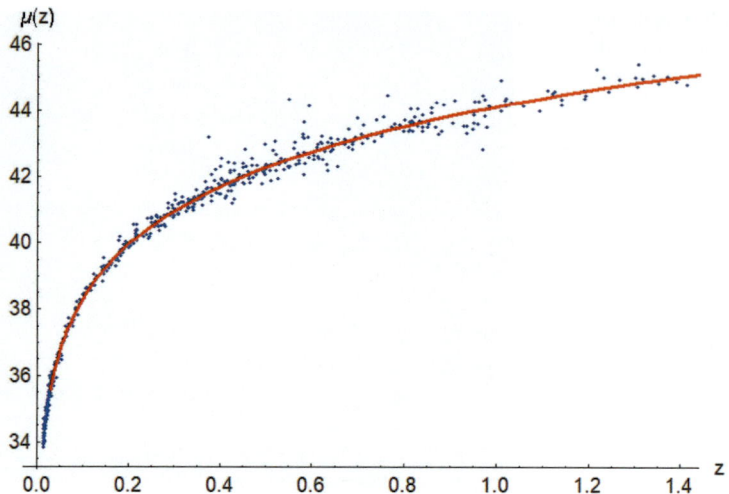

Fig. 7.7 Graph of the distance modulus vs redshift from data of 580 supernovae. (Courtesy Jorge Alfaro and Pablo González, *Universe* 5 (5):96 (2019))

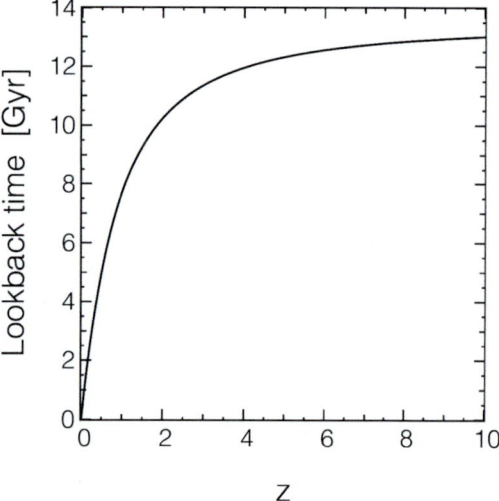

Fig. 7.8 Graph of the lookback time vs redshift z using a standard model of cosmology and Hubble Constant 70.0. Courtesy Georg. Feulner, A Near-Infrared Selected Galaxy Redshift Survey, PhD Dissertation, Munich, 2004

Expansion Scale Factor

There is also an analytical method to derive Hubble's Law by considering what is known as a *scale factor* of the Universe $a(t)$ that depends on the elapsed time t since the Big Bang. Then the distance D_0 between any two observers who live on distant galaxies at a specific time t_0 can be written as $D_0 = D(t_0)$ since the distance is varying with time. This distance variation with time can be expressed in terms of the scaling factor

Expansion Scale Factor

$$D(t) = a(t)D_0 \quad (7.6)$$

which is to say that $a(t) = D(t)/D(t_0)$. The value $a(t_0) = 1$ represents the present-day Universe.

The radial velocity v_r between the two observers is just the rate of change of $D(t)$, so that

$$\begin{aligned} v_r &= \frac{d}{dt}D(t) = \dot{a}(t)D_o \\ &= \frac{\dot{a}(t)}{a(t)}D(t), \end{aligned}$$

where we have substituted the value of D_0 from Eq. (7.6). If we denote

$$H(t) = \dot{a}(t)/a(t),$$

this gives us an expression of the Hubble 'constant' which actually varies over eons of time. The preceding also gives us an elementary derivation of Hubble's Law (Eq. 7.4). We use the notation $H_0 = H(t_0)$ to indicate our present era which is at the time t_0.[4] Note that this is a measure of what is called the *Hubble Flow*, that is the recessional velocity due to the expansion of the Universe. There are also gravitational effects on galaxies from neighboring galaxies which will have some bearing on the recessional velocity but these will not be taken into account. Indeed, our neighboring galaxy, Andromeda, exhibits a blue shift in its light and our Milky Way and Andromeda will eventually form a merger.

It is actually possible to give a heuristic argument as to why galaxies at a further distance are receding faster. Suppose you take an elastic band and you place markings on it at some fixed zero position and then at 1 cm, 2 cm, 3 cm, and 4 cm from left to right. Then pull the band from the right for a duration of 1 s so that the mark at 1 cm has moved 1 cm to the right as measured by a ruler. Then, the point marked at 2 cm on the band will have moved 4 cm to the right, the point at 3 cm will have moved 6 cm to the right and the point at 4 cm will now be 8 cm further to the right. In other words, the first point will be moving away from the origin at 1 cm/s, the second point at 2 cm/s, the third point at 3 cm/s and the fourth point at 4 cm/s. The reader can convince themselves that this actually happens by doing this thought experiment with an elastic band.

---

[4] It should be noted that Edwin Hubble was not the first to discover the expansion of the Universe. Indeed, the Belgian Priest Georges Lemaître had derived the expansion from the General Theory of Relativity, even using observational data to obtain 'Hubble's Law' two years prior to Hubble in 1927.

As one might expect, the scale factor is related to the redshift by the formula

$$\boxed{a(t) = \frac{1}{1+z}}. \tag{7.7}$$

To see why this is so, let λ_e be the wavelength of light when it was emitted, and λ_o be its wavelength when it was observed so that the redshift is

$$z = \frac{\lambda_o - \lambda_e}{\lambda_e},$$

implying that: $\lambda_o = (1 + z)\lambda_e$.

Now, the scale factor $a(t)$ describes how the size of Universe changes over time and as the Universe expands, so the wavelengths of light (radiation) increase proportionally. This is to say that the ratio of the observed wavelength to the emitted wavelength equals the ratio of the scale factors at the time of observation and omission, i.e.

$$\frac{\lambda_o}{\lambda_e} = \frac{a(t_0)}{a(t_e)},$$

where t_0 represents the present, so that $a(t_0) = 1$. Since from above we have found that $\frac{\lambda_o}{\lambda_e} = 1 + z$, we conclude that $1 + z = 1/a(t_e)$ and replacing t_e by any time t gives Eq. (7.7).

For example, $z = 0$ corresponds to the present time with $a(t) = 1$. And if a celestial body has redshift $z = 1$, then $a(t) = 1/2$, meaning that the Universe was one-half its current size and hence has doubled in size since the light was emitted. Likewise, a redshift of $z = 2$ gives $a(t) = 1/3$ and indicates that the Universe has tripled in size since the light was emitted.

One interesting consequence of the Hubble Law is if we put in $v_r = c$ the velocity of light. Then taking, for example, $H_0 = 71$ (km/s)/Mpc, and using Eq. (7.4),

$$D = \frac{299{,}792\,\frac{\text{km}}{\text{s}}}{\left(71\,\frac{\frac{\text{km}}{\text{s}}}{\text{Mpc}}\right)}$$

$$\approx 4222 \text{ Mpc}$$

which is about 13.8 billion light-years. This means that at further distances, namely, $D > c/H_0$, the Universe is expanding faster than light. The value of $D = c/H_0$ is known as the *Hubble radius*. Since the radius of the observable Universe is about 46.5 billion light-years, this means that a significant portion of it consists of regions expanding faster than the speed of light.

Chapter 8
Relativity

Space and time are modes by which we think, not conditions under which we live...
Albert Einstein.

Einstein's Theory of Relativity was developed separately in two parts, the Special Theory published in 1905 dealing with a flat space (that is, without gravity) but including time, and the General Theory published in 1915 that included gravity as well as space and time. An early supporter of Einstein's rather esoteric theories was Sir Arthur Eddington, a scientist at Cambridge University. There is an anecdote (in many versions) that after one of Arthur Eddington's lectures he was queried by physicist Ludwik Silberstein, "Professor Eddington, you must be one of three persons in the world who understands General Relativity." When Eddington paused for a moment, Silverstein replied, "Don't be modest, Eddington", to which Eddington replied, "On the contrary, I am trying to think who the third person is."

If this encounter did indeed happen like this, it was not even true as there was likely a dozen or more physicists at the time who would have understood General Relativity to some extent. However, it makes for a great story and indicates just how arcane Einstein's theories seemingly were at the time. On the other hand, over the intervening decades there have been enumerable explications of both theories and much of it nowadays is accessible as we will demonstrate. Moreover, both aspects of Relativity are invoked on a daily basis by everyday citizens!

Special Relativity

As related in Chap. 3, a person walking on a train in a forward direction at a speed of 5 km/h while the train is travelling at a speed of 100 km/h results in the forward speed of the person relative to the ground being 105 km/h. Likewise if the person walks towards the back of the train at the same speed, their speed relative to the ground will be 95 km/h. But this simple and rather obvious view was about to radically change in the early twentieth century.

The nineteenth century saw a great many developments in Physics. One in particular was the equations of James Clerk Maxwell in 1861–1862. They are a set of equations describing all forms of electromagnetic radiation from gamma rays to radio waves as depicted in Fig. 3.2. But one of the consequences of the equations is that the speed of light is constant in a vacuum.[1] Moreover, the equations implied that the speed of light would remain constant relative to a stationary *luminiferous aether* that permeated all of space. This aether was thought to be necessary because just as sound waves require the medium of air (or water) for their propagation, it was thought at the time that light would require some sort of medium[2] for its propagation as well.

However, just as the Earth is travelling around the Sun at 30 km/s, this would create an 'aether wind' that was expected to affect the speed of light as it travelled at different angles to the wind, such as parallel or transverse to the wind. And this would violate the constancy of the speed of light as formulated by Maxwell.

Not many years later in 1887 two American physicists, A.A. Michelson and E.W. Morley performed a famous experiment that now bears their name in order to compare the speed of light in directions perpendicular to one another so see if the speed of the Earth through the aether wind had any effect on the speed of light (Fig. 8.1). Rotating the device (called an *interferometer*) and taking the measurements at different times of the year again had no effect on the velocity of light. For their efforts, Michelson and Morley won the Nobel Prize in Physics in 1907.

Exercise In the Michelson-Morley experiment (prior to any Relativity considerations) show that if the apparatus is moving with a velocity v (say, due to the Earth's orbital motion) directly into a stationary 'aether', then the expected elapsed time Δt taken for a light beam to travel a distance L and then back the same distance L in the opposite direction (that is, from M to M_2 and back to M) would be given by

$$\Delta t = \frac{2L}{c\left(1 - \frac{v^2}{c^2}\right)},$$

as in Fig. 8.1.

We remark that the elapsed time for the transverse path from M to M_1 and back to M, is essentially discussed in a subsequent section. This is the case where the direction of motion is perpendicular to that of the light beam.

Various explanations of the null result of the Michelson-Morley experiment were proposed such as *aether dragging* whereby the Earth could be 'dragging' the aether

---

[1] This is a consequence derived from Maxwell's Equations that the speed of light c is given by, $c = 1/\sqrt{\epsilon_0 \mu_0}$ where ϵ_0, μ_0 are specific constants relating to a vacuum, namely the *permittivity* and *the permeability* respectively. It is not necessary that we discuss these further.

[2] Generally considered stationary.

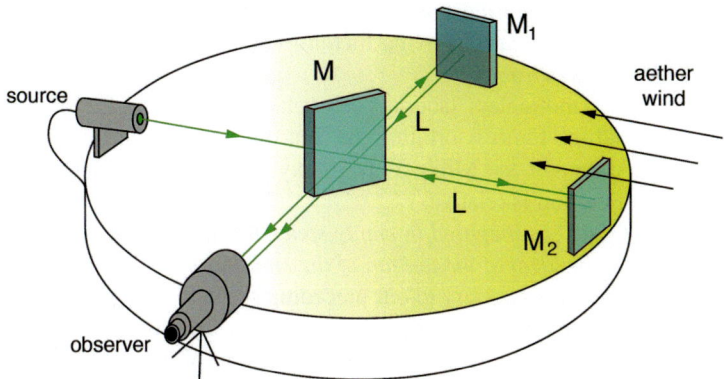

Fig. 8.1 Diagram of the setup of the 1887 Michelson-Morley experiment showing a light beam from a source hitting a partially silvered mirror so that the deflected light beam hits mirror M_1 with the un-reflected beam going straight through to mirror M_2. The two light beams are then recombined at M and sent to an observing eyepiece as indicated. In this scenario, one light-path direction is parallel to the 'aether wind' and the other has a transverse direction. The entire apparatus can be rotated since the direction of the aether wind with respect to the Earth's orbital motion would not be known. (Courtesy Katy Metcalf)

with it, but all were subsequently dismissed. Closer to the mark were explanations of length contraction and time dilation by Lorentz, FitzGerald, and Poincaré which led[3] to their being incorporated into the more general framework of Special Relativity, which did away completely with the notion of a luminiferous aether. But the notions of length contraction and time dilation are indeed real effects.

Two Postulates of Relativity

For the Special Theory of Relativity, Einstein adopted two postulates which he assumed to be to be true.

I. *The laws of physics remain the same in all inertial frames of reference.*
   An *inertial frame of reference*[4] is one that is not undergoing any acceleration, that is, one that is at rest or moving at a constant velocity in a straight line. This could be a train traveling at a constant speed of 200 km/h (with respect to the ground) or a motionless laboratory, and so the laws of Physics are the same for any observer either on the train or in their lab. Inertial frames are in constant rectilinear motion with respect to one another. One gets this sensation if your car

---

[3] There are conflicting views as to what extent, *if any*, the Michelson-Morley experiment affected Einstein's thinking on Special Relativity. Even Einstein himself gave conflicting accounts.

[4] The notion of an inertial frame of reference stems from the work of Galileo on the concept of inertia and is also referred to as a *Galilean reference frame*.

is in a carwash and the rollers are moving from the front to the back of the car, then you feel as if the car is moving forward with respect to the rollers and in a relative sense this is true.

As well, our motionless laboratory is rotating about the Earth's axis once every 24 h and the Earth is orbiting the Sun once a year, and the Sun itself is orbiting the center of the Milky Way galaxy, which itself is moving through space. So, all motion is relative.

II. *The speed of light[5] propagated in empty space has a constant velocity c that is completely independent of the motion of the emitting body.*

This is completely contrary to our preceding scenario of a person walking on a train. But not so with light and the same holds true for all electromagnetic radiation. This was supported by the null result of the Michelson-Morley experiment and the expression for c given by Maxwell.

Time Dilation

Although most people think of time as immutable, that turns out not to be so. Time is affected by motion and also by gravity. Considering motion first, according to the Special Theory of Relativity, velocity affects the rate at which time elapses. As it turns out, a clock traveling in a spaceship at high speed will run slower compared to a stationary clock. It is one of the consequences of the fact that light travels at a constant speed which is postulate II above. Now you will see just how true this is.

In order to keep time in our moving frame of reference such as a spaceship,[6] we will use a *light clock* as shown in Fig. 8.2. This is simply a thought experiment, but has the virtue of avoiding the need to consider gears or friction found in ordinary clocks. Moreover, the speed of light is constant in all inertial reference frames and that is a crucial factor here. The light clock consists of two small mirrors located at positions A and B at a distance d between them, which bounce a beam of light back and forth between them. One 'tick' of the clock consists of one round trip from A to B and back to A again and this time interval for the 'stationary' clock will be denoted by Δt_0. If we think of the clock sitting on the surface of the Earth, we have to then ignore the effects of gravity discussed above, so let us do just that and assume that there are no gravitational effects acting on the clock.

Now consider an identical clock on a spaceship moving at a constant velocity, say v, and let us measure the time duration of a tick in this scenario. In this instance, the light path will take a zigzag path relative to a stationary observer on the ground. The distance travelled for a tick of the clock on the spaceship will be $2D$ versus the duration $2d$ of a tick of the stationary clock and clearly $D > d$. But how much greater depends on the velocity of the spaceship (see Fig. 8.3).

---

[5] It will always be assumed that the speed of light c is being measured in a vacuum.

[6] Our spaceships will always be traveling at a constant velocity with respect to their frame of reference.

Time Dilation

Fig. 8.2 Our mental light clock where one 'tick' of the clock is taken as the time for a light beam to go from A to B and back to A again. (Courtesy Katy Metcalf)

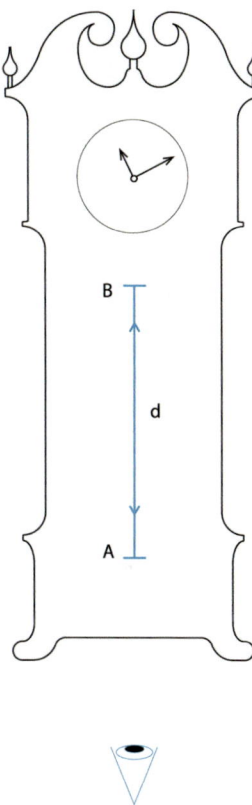

In order to compare the rates at which the two clocks measure time, we simply require the basic formula *distance = velocity × time*, and since the velocity is the velocity of light c, we have the general expression[7]

$$d = c\Delta t.$$

Let us denote the time duration of a tick of the stationary clock as Δt_0 to differentiate it from the time duration of a tick of the moving clock aboard the spaceship (Fig. 8.3), which we denote by Δt. As in each case the velocity is c, we can solve for the distance travelled by a half-tick for both the stationary clock and the one aboard the spaceship, namely

---

[7] Normally, this equation is written simply as $d = v \times t$, but we wish to emphasize that the time t represents the elapsed time so we write it as Δt.

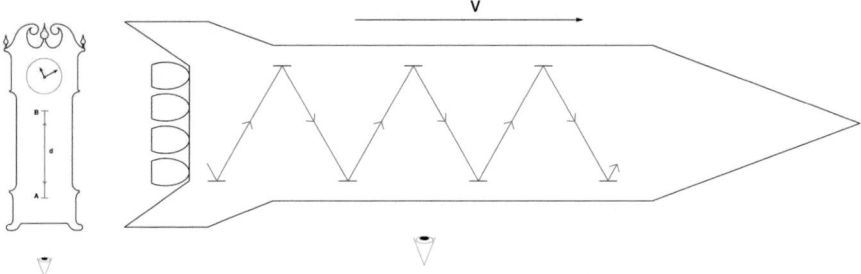

Fig. 8.3 On the spaceship the light in our light clock now takes a zigzag path as seen from a stationary observer and thus each 'tick' of the clock will have longer duration than a stationary clock. Thus, by comparison, time will slow down for spaceship crew members. (Courtesy Katy Metcalf)

$$d = \frac{c\Delta t_0}{2},$$
$$D = \frac{c\Delta t}{2}.$$

We can now put these two values to use in order to compare them as in Fig. 8.4 where the base of the triangle formed by the two distances is denoted by a.

The two distances d and D are related by the Pythagorean theorem,

$$D^2 = a^2 + d^2.$$

The distance a is determined by the velocity v of the rocket, that is, $a = v\frac{\Delta t}{2}$, so that the preceding equation becomes,

$$\left(\frac{c\Delta t}{2}\right)^2 = D^2 = a^2 + d^2$$
$$= v^2\left(\frac{\Delta t}{2}\right)^2 + \left(\frac{c\Delta t_0}{2}\right)^2$$
$$= \frac{v^2 \Delta t^2}{4} + \frac{c^2 \Delta t_0^2}{4}.$$

We cancel the 4 on both sides and rearranging the terms leads to

$$\Delta t^2 \left(c^2 - v^2\right) = c^2 \Delta t_0^2.$$

If we now divide both sides by c^2 we have

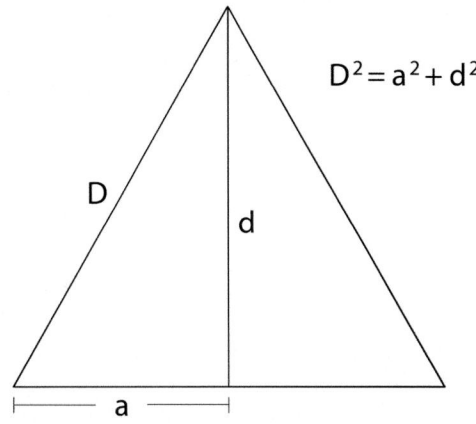

Fig. 8.4 The relation between the distance D for half-tick of the clock on board the spaceship compared to the distance d of half-tick of the stationary clock. (Courtesy Katy Metcalf)

$$\Delta t^2 \left(1 - \frac{v^2}{c^2}\right) = \Delta t_0^2.$$

In other words,

$$\boxed{\Delta t = \frac{\Delta t_0}{\sqrt{1 - \frac{v^2}{c^2}}}}. \tag{8.1}$$

which gives the relation between the two time intervals.

Exercise Using the above relativistic approach replacing d by L for a light beam that is travelling perpendicular to the direction of motion of velocity v, show that the total travel time $\Delta t'$ of a light beam in the Michelson-Morley experiment from M to M_1 and back to M is given by

$$\Delta t' = \frac{2L}{c\sqrt{1 - \frac{v^2}{c^2}}}.$$

Rotating the interferometer in all different directions produced no noticeable observable effects of the combined light beams at the observer eyepiece. Compare this with the preceding value determined in the non-relativistic scenario.

Lorentz Dilation Factor

Here we see that the (*Lorentz*) *time dilation factor* is given by

$$\frac{1}{\sqrt{1-\frac{v^2}{c^2}}},$$

and shows how the time duration Δt of a tick of a clock moving at velocity v is to be compared to that of a tick of a stationary clock, Δt_0.[8]

Note that since the denominator of the dilation factor is less than or equal to 1, so that $\Delta t \geq \Delta t_0$ and there is only equality when $v = 0$, that is, when the spaceship is also stationary. If the spaceship has any velocity at all, then $\Delta t > \Delta t_0$ and hence time passes more slowly on the spaceship. Time has dilated on the moving spaceship and the amount of dilation depends on the velocity it is moving at. The time dilation factor will be greater with higher the velocity v as then the denominator becomes even smaller.

For example, if someone is aboard a spaceship traveling at 98% the speed of light, then $v = 0.98c$ and the time dilation factor is

$$\frac{1}{\sqrt{1-\frac{v^2}{c^2}}} = \frac{1}{\sqrt{1-(0.98)^2}} \approx 5,$$

implying that $\Delta t \approx 5\Delta t_0$. This means that the tick of a clock onboard the spaceship has 5 times the duration relative to the tick of a stationary clock and hence the elapsed time aboard the spaceship will be one-fifth of that relative to stationary time.

Exercise Show that for someone on a spaceship travelling at a speed of 99.5% the speed of light, when they have experienced 1 year of time having elapsed, then their identical twin on the ground will have experienced that 10 years of time has elapsed. Such considerations became known as the 'twin paradox' and is rather mysterious, but the outcome is valid.

It must be said that the twin paradox produced a profound sense of disbelief in many scientists and philosophers. Everyone agreed that the mathematics was correct but the conclusion could not possibly apply to the real world of the living. This produced decades of consternation in books and debates, between those who believed it most definitely applied to the real world (including biological systems), and those who vehemently opposed such views. Of course, Einstein himself waded into the fray as did many of his supporters but the issue is now largely settled as we

---

[8] This same time dilation factor was invoked by Lorentz to partially explain the result of the Michelson-Morley experiment.

shall see in the sequel. However, understanding the metaphysical nature of 'time' itself, is one of ongoing relevance.

Interestingly, the initial mass m_0 of an object moving at a constant velocity v will also experience a similar *mass dilation factor* in that its mass will become increased to m given by

$$\boxed{m = \frac{m_0}{\sqrt{1 - v^2/c^2}}}. \qquad (8.2)$$

Clearly, when $v = 0$ we have $m = m_0$, the rest mass, and that m will increase as v approaches c. This is an important factor in particle accelerators that actually do achieve speeds close to that of c. There is an analogous length contraction formula applicable in the direction of motion.

General Relativity

Some 10 years after the publication of Einstein's Special Theory of Relativity which did not include gravity, came the publication of the profound General Theory of Relativity which did incorporate gravity and revolutionized the study of how the Universe works. This 10-year span represents one of the greatest scientific achievements in human history, evolving over a number of papers that culminated in the General Theory.

One of the fundamental concepts of the General Theory is the *Equivalence Principle* introduced in 1908[9] but not under a specific name:

> ... in the discussion that follows, we shall therefore assume the complete physical equivalence of a gravitational field and a corresponding acceleration of the reference system.

Einstein called this: "The happiest thought of my life."

In a thought experiment one can imagine a person in an elevator that is being accelerated upward and who will experience an attraction no different from the way a person experiences gravity on Earth as in Fig. 8.5. If the acceleration is equal to 1 g then the experiences will be exactly the same.

---

[9] On the relativity principle and the conclusions drawn from it (1908) *Jahrbuch der Radioaktivitaet und Elektronik* 4, p.443.

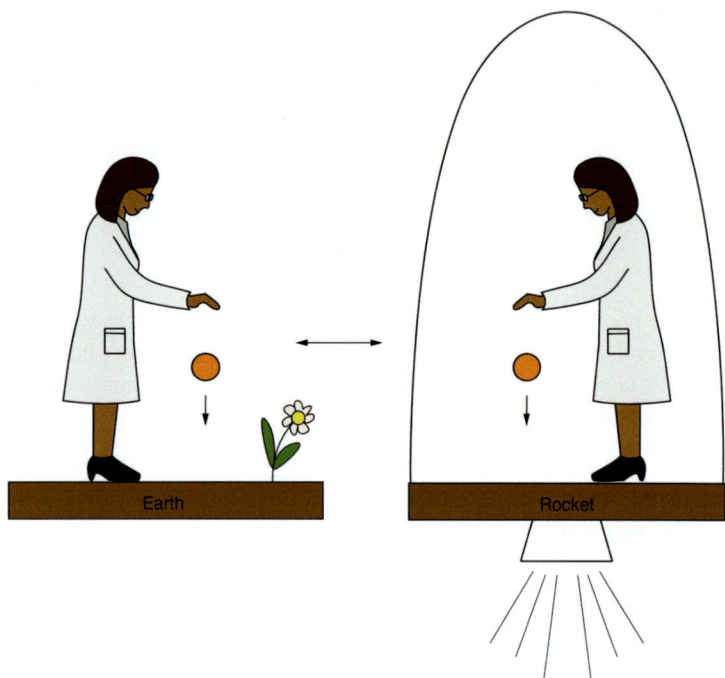

Fig. 8.5 The equivalence of gravity and acceleration whereby a person in an elevator being accelerated at 1 g will find that an object falls at the same rate as it does on Earth and the person will experience their own weight as the same. (Courtesy Katy Metcalf)

Time Dilation

Considering now the matter of gravity, if we have a clock on the surface of a spherical body of mass M and another clock at a very distant point not under the gravitational influence of any body then the *time dilation factor* is in accordance with the General Theory of Relativity[10]

$$\Delta t_g = \Delta t_0 \sqrt{1 - \frac{2GM}{Rc^2}}, \qquad (8.3)$$

where Δt_0 is the elapsed time of a distant observer not under the gravitational influence of the body, Δt_g is the corresponding elapsed time on the surface of the

---

[10] In this section we are not assuming any time effects due to velocity.

body at a radius R from the center and G and c are the gravitational constant and velocity of light respectively. Observe that this means that $\Delta t_g < \Delta t_0$, which is to say that time runs slower under the influence of a gravitational field.

We will see in Chap. 9 that the quantity

$$\frac{2GM}{c^2}$$

represents the (Schwarzschild) radius, R_S, of a body of mass M at which the body becomes a black hole and no light can escape it. Thus, we could rewrite Eq. (8.3) as

$$\Delta t_g = \Delta t_0 \sqrt{1 - \frac{R_S}{R}},$$

and from this we can deduce that we require the $R > R_S$ for otherwise we encounter the square root of a negative number and hence imaginary time which can be problematic. But this is just a technical proviso and we need not worry about it.

The mass of the Earth does affect the time on the surface as we will discuss below but it is by a quite small amount and not easily determined on a calculator so let us consider a more massive body like the Sun whose mass is $M_\odot = 1.9885 \times 10^{30}$ kg and radius $R_\odot = 696{,}340$ km. Even at this mass the time dilation factor will be small so we will use all available significant figures at our disposal. Therefore, by Eq. (8.3) and without the units which all cancel out anyway,

$$\Delta t_g = \Delta t_0 \sqrt{1 - \frac{2GM}{Rc^2}},$$
$$= \Delta t_0 \sqrt{1 - \frac{2 \times (6.67430 \times 10^{-11}) \times (1.9885 \times 10^{30})}{(6.96340 \times 10^8) \times (299{,}792{,}458)^2}},$$
$$= \Delta t_0 \times 0.99999788,$$

according to a short computer program written by the author. If we put in a year's worth of seconds: $\Delta t_0 = 31{,}557{,}600$ s, then we obtain the elapsed time on the surface of the Sun as $\Delta t = 31{,}557{,}533$ s, so that the time difference is 67 s slower per year on the surface of the Sun. See Appendix VII for various Python programs related to time dilation.

Exercise Calculate the time difference over a period of one year between a very distant observer in space compared to the time at the surface of the Earth. *Ans.* 0.022 s/year, computed by the author's Python program in Appendix VII.

Global Positioning System

Thus far, what we have done regarding time dilation as a consequence of gravity and velocity has been rather theoretical and you might think of little consequence to the average person in the street. Wrong, in view of the Global Positioning System (GPS) that is used every day for personal and business navigation. The system consists of an array of 31 satellites in an orbit of 20,200 km altitude around the Earth with each having an onboard atomic clock that keeps very precise time. This guarantees a minimum of four satellites that are in view of a ground observer at any one time.

Given a single satellite A that is a distance a from say you, who is lost, means that you are located somewhere on a sphere having center A and radius a. Likewise, if you are distances b from satellite B and distance c from satellite C again leads to your position being on two spheres centered at B and C with radii b and c respectively. Then the intersection of these three spheres will result in two distinct points, but only one of them will be on the Earth's surface. This is your latitude and longitude on Earth. By incorporating a fourth satellite, your altitude can also be ascertained in a process known as *trilateration*.

But how are the distances from each satellite measured? This is done by the broadcast of a radio signal travelling at the speed of light with a precise timestamp of when it was sent and its precise location in space at that moment. A GPS unit of an individual on the ground does not have an atomic clock but is automatically updated with atomic clocks in order to obtain the necessary precision for your position location. Then, by measuring the time difference Δt between the time registered by the timestamp from each satellite and the time received, then $d = c\Delta t$ determines the distance between the person on the ground and each satellite.[11] Hence your position on Earth is now located.

There is just one problem. To calculate the time difference Δt between when a satellite signal was sent and when it was received, the atomic clocks in the satellites have to be synchronized with atomic clock time on the ground. Unfortunately, as we have seen in the preceding sections, both velocity and gravity affect the keeping of time.

The atomic clocks aboard the satellites are orbiting at some 14,000 km/h with a consequence that they will be running about 7.3 millionths of a second per day slower than a stationary clock. On the other hand, being at an altitude of 20,200 km means that the clocks in the satellites experiencing a lesser gravitational pull than Earth based clocks and so would be running approximately 45.8 millionths of a second per day faster on this account.

These two opposite effects are not equal so that overall, the satellites' atomic clocks would run about 38.5 millionths of a second per day faster than Earth based atomic clocks. This means that before the atomic clocks are launched into space,

---

[11] The Earth's atmosphere can affect the speed of radio transmission somewhat, but these will not be discussed here.

they must be tweaked to slow down by just the right amount to run in sync once in orbit with the atomic clocks on Earth.

Exercise Simple computer programs might be required for this exercise. They can also be found in the Appendix VII.

(a) Using Eq. (8.3) show that a clock on the Earth's surface slows down by 6.02×10^{-5} seconds per day due to gravity compared to a clock in deep space.
(b) Show that a GPS satellite slows down due to the Earth's gravity by 1.44×10^{-5} seconds per day compared to a clock in deep space.
(c) Show then the *net* time gain due to gravity of a clock aboard the GPS satellite compared to an Earth clock is: 45.8 millionths of a second per day.[12]
(d) Show that a GPS satellite with a velocity of 14,000 km/h will be running about 7.3 millionths of a second per day slower than a stationary clock.

Apsidal Precession

Due to General Relativity, the orbit of a body about a large mass will *precess* due to the warping of spacetime. One of the first affirmations of Einstein's General Theory of Relativity was accounting for the precession of the orbit of Mercury (Fig. 8.6 (L)). This orbit (that is, the *apsis/apsides*) has been known to gradually rotate (precess) and the amount of rotation due to other planets came up short by 43 arcseconds per

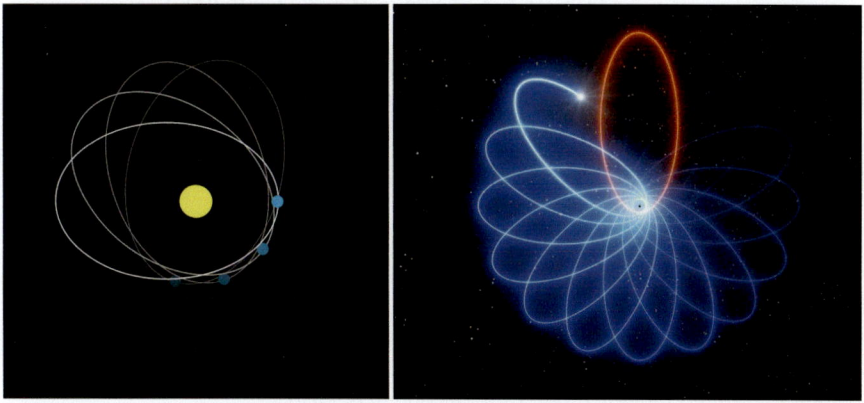

Fig. 8.6 (L) Exaggerated precession of the orbit of Mercury about the Sun; (R) Artist impression of the precession of the star S2 about the central black hole of the Milky Way Sgr A* also exaggerated. (Courtesy (L) Public Domain / Rainer Zenz; (R) ESO/L. Calçada)

---

[12] If Earth − deep space = A; Satellite − deep space = B; then Earth − Satellite = A − B.

century.[13] A hypothetical planet named Vulcan orbiting between Mercury and the Sun was proposed to account for this discrepancy although careful astronomical observations subsequently proved this not to be the case.

What accounts for the discrepancy is the following equation derived from General Relativity (see Appendix VIII for Einstein's original published calculation)

$$\boxed{\delta\phi = \frac{24\pi^3 a^2}{T^2 c^2 (1-e^2)}}, \tag{8.4}$$

where a is the length of the semi-major axis in meters, T is the period of rotation in seconds, and e is the eccentricity of the orbit, and the value of $\delta\phi$, the *Schwarzschild (apsidal) precession*, is in *radians per orbit*. In the case of Mercury as depicted in Fig. 8.6 (L), the semi-major axis is

$$a = 5.79 \times 10^{10} \text{ m};$$
$$T = 87.97 \text{ d} \times 86400 \text{ s/d} = 7.60 \times 10^6 \text{ s};$$
$$e = 0.206;$$
$$c = 2.998 \times 10^8 \text{ m/s}.$$

Then we have

$$\delta\phi = \frac{24\pi^3 \times (5.79 \times 10^{10})^2}{(7.60 \times 10^6)^2 \times (2.998 \times 10^8)^2 \times (1 - 0.206^2)}$$
$$= 5.02 \times \frac{10^{-7} \text{rad}}{\text{orbit}}.$$

Since each orbit is 87.97 days and one century is 36524 days, then in one century Mercury has gone 36524/87.97 = 415.19 orbits. We conclude that

$$\frac{\delta\phi}{\text{century}} = 5.02 \times 10^{-7} \frac{\text{radians}}{\text{orbit}} \times 415.19 \text{ orbits}$$
$$= 2.08 \times 10^{-4} \text{rad}$$
$$= 119.18 \times 10^{-4} \text{deg},$$

having multiplied radians times $180/\pi$ to obtain degrees.

Finally, we need to convert this to arcseconds and as there are 3,600 arcseconds per degree, we have

---

[13] There are other factors of small significance affecting the precession which will not be discussed. The total apsidal precession of the orbit of Mercury is 5.75 arcsec per year, or 575 arcsec/century. Of this, 532 arcsec/century are accounted for by the gravitational attraction of other bodies in the Solar System.

Apsidal Precession

$$\frac{\delta\phi}{\text{century}} = (119.18 \times 10^{-4} \text{ deg}) \times 3600 \, \frac{\text{arcsec}}{\text{deg}}$$

and so

$$\frac{\delta\phi}{\text{century}} = 42.9 \text{ arcsec}.$$

Thus, the precession of the orbit of Mercury due to relativistic effects is about 43 arcseconds per century and accounts for the discrepancy initially observed.

The Earth's orbit also exhibits precession and it takes 112,000 years for the orbit to make a complete 360° rotation back to its initial orientation in space. Again, this is a combination of various effects including that given by General Relativity as above.

We can solve for the period of rotation via Eq. (5.15) for a large body of mass M being orbited by a much smaller body which gives

$$T^2 = \frac{4\pi^2 a^3}{GM}.$$

Putting this expression into the formula Eq. (8.4) results in an equivalent formulation

$$\boxed{\delta\phi = \frac{6\pi GM}{c^2 a \, (1-e^2)}} \qquad (8.5)$$

for Schwarzschild precession.

For example, let us work out the Schwarzschild precession of the star S2 that is orbiting the Milky Way central black hole Sgr A* as depicted in Fig. 8.6 (R). We can use either Eq. (8.4) or Eq. (8.5) so let us use the former again as it involves slightly less computation. For the constants the semi-major axis was worked out in Exercise* of Chap. 5 as

$$a = 15.35 \times 10^{13} \text{m};$$

$$T = 16.05 \text{ year} = 16.05 \text{ year} \times (3.16 \times 10^7 \text{s/year}) = 50.72 \times 10^7 \text{ s};$$

$$c = 2.998 \times 10^8 \text{m/s};$$

$$e = 0.885.$$

Putting these values into Eq. (8.4)

$$\delta\phi = \frac{24\pi^3 a^2}{T^2 c^2 (1-e^2)} = \frac{24\pi^3 \times (15.35 \times 10^{13} \text{ m})^2}{(50.72 \times 10^7 \text{ s})^2 \times (2.998 \times 10^8 \text{ m/s})^2 (1 - 0.885^2)}$$

$$\frac{744.15 \times 235.62 \times 10^{26}}{(2572.5 \times 10^{14})(8.99 \times 10^{16})(0.217)} = 34.9 \times 10^{-4} \text{ rad}.$$

Converting radians to degrees, we obtain

$$\delta\phi = 34.9 \times 10^{-4} \text{ rad} \times \frac{180}{\pi} \approx 0.2 \text{ deg}.$$

As there are 60 arcmin in each degree we arrive at

$$\boldsymbol{\delta\phi \approx 12 \text{ arcmin}}$$

per orbit. This result of General Relativity is in complete accord with direct observation of the apsis precession.[14]

Exercise As we have already computed the mass of Sgr A* as $M = 4.14 \times 10^6 M_\odot$ previously, use Eq. (8.5) to compute the Schwarzschild precession of the star S2.

Gravitational Redshift

A gravitational field affects spacetime, mass, and indeed all electromagnetic radiation. In the latter case, this is due to the fact that the waves lose energy due to them having to climb out of a 'gravity well'. Since the energy of the waves is given by $E = hf$, and the relation between frequency and wavelength is $\lambda = c/f$, we observe that as the energy decreases, the wavelength *increases*, becoming redder. This increase in wavelength is known as the *gravitational redshift* and is completely distinct from the redshift due to recessional velocity (Fig. 8.7).

As it turns out the gravitational redshift is proportional to the mass M of a star divided by the radius R, that is

$$z \propto \frac{M}{R}.$$

A simple formula derived from the General Theory of Relativity is given in terms of frequency shift is for small redshift values

---

[14] R. Abuter et al., Detection of the Schwarzschild precession in the orbit of the star S2 near the Galactic centre massive black hole, *A&A*, 636, id. L5 (2020), 14 pp. The authors also measured ≈12 arcmin per orbit.

Gravitational Redshift

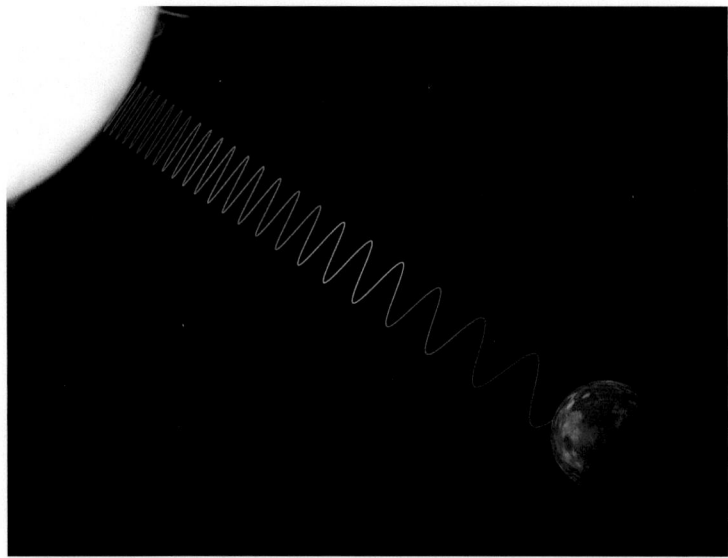

Fig. 8.7 Depiction of the increase in wavelength of light due to the gravitational effect of the mass of the star. (Courtesy Jonathan Park)

$$z = \frac{f_e - f_o}{f_e} = \frac{GM}{Rc^2}. \tag{8.6}$$

where in this instance, f_e is the frequency of light emitted from the surface of the star and f_o is the observed frequency from a large distance.

The Eq. (8.6) is fine, but astronomers usually work in units of solar masses and solar radii and so this formula can be simplified even further. Indeed, the simplified expression is given in terms of cz which astronomers prefer, that is,

$$v_{gr} = cz = 0.636 \frac{M_\odot}{R_\odot}, \tag{8.7}$$

where $M_\odot$ is measured in solar masses and $R_\odot$ represents solar radii.[15]

To see that Eq. (8.7) is effectively Eq. (8.6), note that the latter equation is (without units)

$$cz = \frac{GM}{Rc} = \frac{(6.674 \times 10^{-11})}{2.998 \times 10^{11}} \times \frac{M}{R} = (2.226 \times 10^{-22}) \frac{M}{R}.$$

On the other hand, for Eq. (8.7), converting M and R to $M_\odot$ and $R_\odot$ respectively gives,

---

[15] For reference, the radius of the Sun is 6.963×10^8 meters.

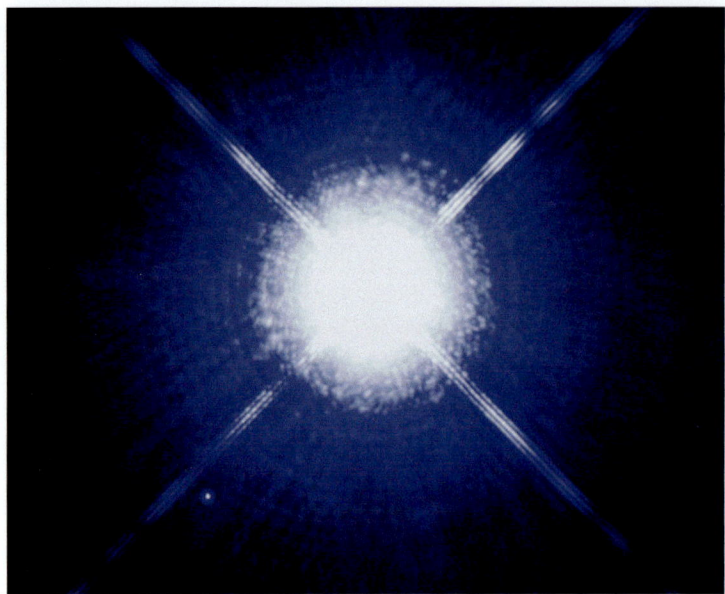

Fig. 8.8 Sirius A with its companion white dwarf (bottom left). (Courtesy NASA, ESA, H. Bond (STScI), and M. Barstow (University of Leicester). Courtesy NASA/Hubble Space Telescope)

$$cz = 0.636 \frac{M_\odot}{R_\odot} = 0.636 \times \left(\frac{M}{1.989 \times 10^{30}}\right) / \left(\frac{R}{6.963 \times 10^8}\right) = \left(2.226 \times 10^{-22}\right) \frac{M}{R}.$$

Thus, we have equality to three decimal places and that is good enough for us.

The gravitational redshift effect then will be most observable whenever the mass of the star is large and its radius is small, that is in white dwarfs. When the white dwarf has a much larger companion, the recessional velocity redshift can be taken into account via the companion star.

One such nearby example is Sirius B, the companion star to Sirius A, at distances 8.7 and 8.6 light-years respectively (Fig. 8.8). For Sirius B, its mass is measured at $M = 1.018\, M_\odot$ and its radius is $R = 0.00803\, R_\odot$. Therefore, the gravitational redshift given by General Relativity is

$$cz = 0.636 \frac{1.018}{0.00803} = 80.63 \text{ km/s}.$$

The measured gravitational redshift is $cz = 80.65$ km/s.[16]

Moreover, from Eq. (8.7) one is able to compute the mass of the star once the gravitational redshift cz and the star's radius are known.

---

[16] Sirius B data taken from: S.R.G. Joyce et al., The gravitational redshift of Sirius B, *MNRAS*, 481 (2018), 2361–2370.

Exercise Another nearby white dwarf (distance $=16.3$ light-years) is 40 Eridani B. Its mass has been determined as $M = 0.573\,M_\odot$ and its radius $R = 0.01308\,R_\odot$.[17] Compare its gravitational redshift with the measured value of 26.5 km/s. It should be noted that the gravitational redshift must be teased out of the recessional redshift discussed earlier. This is achievable in a binary star system as the recessional velocity of the companion can be determined separately. (*Ans.* ~27.86 km/s).

General Gravitational Redshift Formula
The actual more precise formula from General Relativity for all redshift values is

$$\boxed{f_o = f_e \sqrt{1 - \frac{2GM}{rc^2}}}. \tag{8.8}$$

However, we can simplify this expression if we use the approximation $(1-x)^{1/2} \approx 1 - \frac{x}{2}$ for $x \ll 1$.[18] Thus we can write

$$f_o = f_e \left(1 - \frac{GM}{rc^2}\right).$$

This leads to the evaluation of the gravitational redshift as

$$z = \frac{f_e - f_o}{f_e} = \frac{f_e - f_e\left(1 - \frac{GM}{rc^2}\right)}{f_e} = 1 - \left(1 - \frac{GM}{rc^2}\right)$$

$$= \frac{GM}{rc^2},$$

and $v_{gr} = cz = GM/rc$ which is the simplified version of Eq. (8.6).

Chandrasekhar Limit
Since we are dealing with white dwarfs, let us see how massive a progenitor star can be in order to collapse into a white dwarf. Once a star up to about eight solar masses has exhausted all of its hydrogen fuel in fusion to helium it then expands into a red giant sustained by the nuclear fusion of helium into heavier elements such as carbon and oxygen. Eventually, the outer layers of the core are shed forming a planetary nebula and leaving a dense hot core of a white dwarf consisting mainly of carbon and oxygen with no further fusion taking place. Further collapse of the white dwarf is

---

[17] From: H.E. Bond *et al.*, Astrophysical implications of a new dynamical mass for the nearby white dwarf 40 Eridani B, *A J*, 848 (2017), 6 pp.

[18] This comes from approximating the curve $y = (1-x)^{1/2}$ with the tangent line which has a slope of $-1/2$ at $x = 0$, i.e., $y - 1 = \left(-\frac{1}{2}\right)(x - 0)$ and so $y = 1 - \frac{x}{2}$.

prevented by *electron degeneracy pressure* which is a resistance to the further compression of densely packed electrons.[19]

The mass of the core forming a white dwarf can only be of a limiting mass since if it is any larger, then the core will collapse further into a neutron star or black hole. Determination of this limiting mass is at the intersection of Physics, Cosmology, and Relativity and is given by the *Chandrasekhar Limit*

$$\boxed{M_{Chand} = \frac{\omega_3^0 \sqrt{3\pi}}{2} \left(\frac{\hbar c}{G}\right)^{\frac{3}{2}} \cdot \frac{1}{(\mu_e m_H)^2}.} \qquad (8.9)$$

where $\omega_3^0 = 2.018\,236$ is a dimensionless constant, $\hbar$ is the reduced Planck constant, c the velocity of light, G the gravitational constant, and $\mu_e = 2$ is the average molecular weight of the star's atoms per electron,[20] and $m_H = 1.6735575 \times 10^{-27}$ kg is the mass of the hydrogen atom.

The quantity

$$m_P = \left(\frac{\hbar c}{G}\right)^{1/2}$$

is known as the *Planck mass* and its value is

$$m_P = \left(\frac{\hbar c}{G}\right)^{1/2} = \left(\frac{(1.054571817 \times 10^{-34} \text{ kg} \cdot \text{m}^2/\text{s}) \times (2.99792458 \times 10^8 \text{ m/s})}{6.67430 \times 10^{-11} \text{ m}^3/\text{kg} \cdot \text{s}^2}\right)^{1/2}$$
$$= \sqrt{4.736866 \times 10^{-16} \text{ kg}^2}$$
$$= 2.176434 \times 10^{-8} \text{ kg}.$$

Therefore,

$$m_P^3 = \left(\frac{\hbar c}{G}\right)^{3/2} = 10.309 \times 10^{-24} \text{ kg}^3.$$

Putting in all the ingredients into Eq. (8.9) and rounding off,

---

[19] This is due to the *Pauli Exclusion Principle* which states that no two identical fermions, such as electrons, can occupy the same quantum state simultaneously. Thus, when electrons are highly compressed together, they recoil from one another creating a pressure resistant to further compression.

[20] Since the white dwarf is mostly carbon and oxygen, then the average molecular weight per electron of both is: $\frac{12}{6} = \frac{16}{8} = 2 = \mu_e$.

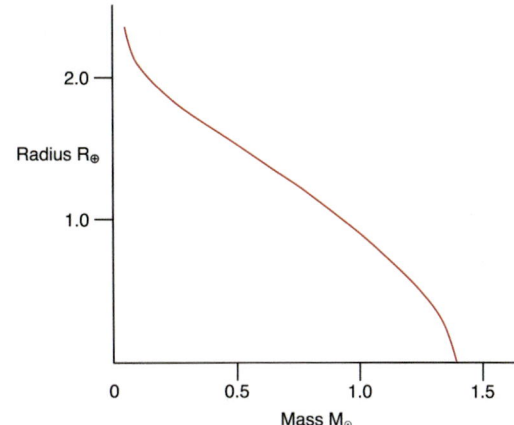

Fig. 8.9 Graph of the mass-radius relationship for white dwarf stars. Note that the radius is given in terms of the Earth's radius and mass in solar masses. The smaller mass white dwarfs are larger in size and if the mass exceeds 1.4 solar masses, the radius size falls to zero in accordance with the Chandrasekhar limit. (Courtesy Aaron Schiff)

$$M_{Chand} = \frac{(2.018)(3.07)}{2} \times \frac{10.309 \times 10^{-24} \text{ kg}^3}{(2 \times 1.674 \times 10^{-27})^2}$$

$$= 2.849 \times 10^{30} \text{ kg}.$$

In terms of solar masses, this amounts to (see Fig. 8.9)

$$M_{Chand} = \frac{2.849 \times 10^{30} \text{ kg}}{1.989 \times 10^{30} \text{ kg}} \approx 1.4 M_\odot.$$

We also conclude that

$$M_{Chand} \propto \frac{m_P^3}{m_H^2}.$$

Gravity and Light

When Einstein's Theory of General Relativity was published in 1915, it resolved the issue of the precession of orbit of Mercury. But it is quite another matter to make a prediction of some cosmic behavior that no one had ever verified previously. One of the predictions of General Relativity was that a massive body such as the Sun would bend light rays as they passed the Sun's rim due to the warping of spacetime as in Fig. 8.10. Fortuitously, in 1919 on May 29th there was to be a total eclipse of the Sun and an expedition to South Africa was undertaken by English astrophysicist Arthur Eddington to observe the eclipse from the island of Principe in the Gulf of Guinea. In addition, another group under the direction of the Royal Astronomer, Sir Frank Watson Dyson, went to observe the eclipse from Sobral, Northern Brazil.

The bending of light waves can be seen in the context of the equivalence of acceleration and gravity (Equivalence Principle). Once again, we can image a person in an elevator accelerating upwards and a beam of light passing through one of its

Fig. 8.10 The bending of light rays owing to acceleration will be the same in a gravitational field in accordance with the equivalence of acceleration and gravity. (Courtesy Katy Metcalf)

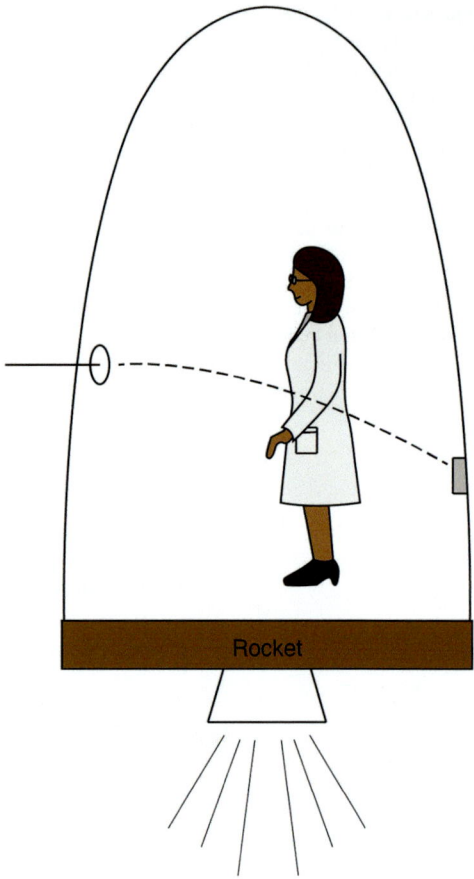

sides. By the time it reaches the other side of the elevator it will hit the opposite wall at a position slightly below the point opposite where it entered. Since acceleration is equivalent to gravity the same should happen in a gravitational field, and it does.

According to Newton, all bodies are affected by gravity, regardless of their mass, even those without mass. He proposed that light was composed of 'corpuscles' that are weightless but would nevertheless be deflected by gravity. The amount of this deflection of light from a distant star close to the rim of the Sun would amount to 0.87 arcseconds. The displacement of a star's light by the Sun as predicted by Einstein would be 1.75 arcseconds, or double the Newtonian amount. It was also possible that both theories were incorrect and that there would be no displacement. So, the test during the 1919 solar eclipse was an important scientific undertaking.

The deflection angle as in Fig. 8.11 according to the General Theory of Relativity is given by

Gravitational Redshift

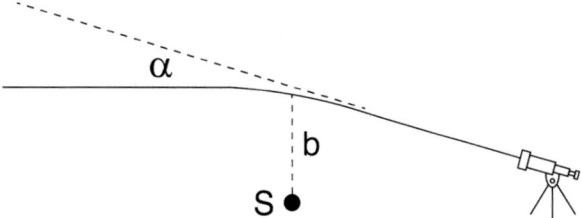

Fig. 8.11 Illustration showing the bending of the light rays due to the warping of spacetime by a mass M. The angle α is the angular deflection from the natural course of the light ray passing at a distance b from a (point) mass M that the ray passes in proximity. (Courtesy Katy Metcalf)

$$\boxed{\alpha = \frac{4GM}{c^2 b}}. \tag{8.10}$$

In the case of the Sun, $M = M_\odot$ and if we observe light rays that have just grazed the Sun's rim, then $b = R_\odot$ since we measure b from the center of the mass M. Therefore,

$$\alpha = \frac{4GM_\odot}{c^2 R_\odot}$$
$$= \frac{4 \times 6.6743 \times 10^{-11} \times 1.989 \times 10^{30}}{(2.998 \times 10^8)^2 \times 6.9634 \times 10^8}$$
$$= 8.484 \times 10^{-6} \text{ rad}.$$

Since there are 206,265 arcsec per radian, we have

$$8.484 \times 10^{-6} \text{ rad} \times \frac{206,265 \text{ arcsec}}{\text{rad}} \approx 1.75 \text{ arcsec}.$$

The so-called 'Newtonian' deflection value is one-half the General Relativity value: $\alpha = \frac{2GM_\odot}{c^2 R_\odot} \approx 0.87$. [21]

The results from both Eddington and Dyson were closer to the higher value although there was, as to be expected, a certain amount of error intrinsic to both

---

[21] Einstein did this very same calculation in his 1911 paper based on his Special Theory of Relativity, in particular gravitational time dilation (*Über den Einfluß der Schwerkraft auf die Ausbreitung des Lichtes*). He determined the (rounded-off) deflection value to be $\alpha = 4 \times 10^{-6}$ **rad = 0.83 arcsec**, based on the constant values available at the time. The paper concludes with the plea: 'It is urgently desirable that astronomers concern themselves with the question brought up here, even if the foregoing considerations might seem insufficiently founded or even adventurous.' On the other hand, Johann Georg von Soldner in 1801 had already obtained a value of **0.84 arcsec** based on Newton's Law of Gravitation and corpuscular theory of light which was quite a remarkable achievement for the time. It was only with Einstein's General Theory of Relativity in 1915 that this displacement value of α was revised by him to **1.75 arcsec** via Eq. (8.10).

measurements. Eddington's results were a bit under the relativistic value (1.61″ ± 0.30″) and Dyson's were a bit over (1.98″ ± 0.30″). As a consequence, the prediction was considered validated,[22] although over the decades there have been some authors who disputed that Eddington and Dyson's results constituted a validation of General Relativity, reducing their results to the status of scientific folklore. But a careful examination of the results in modern times has refuted those assertions and the confirmation of 1919 stands.[23]

Gravitational Microlensing

The lensing effect of one star by another star was a consideration of another short note of Einstein's in 1936.[24] In this instance he discussed the light of source star S passing through the gravitational field of a lens star L such that an observer O is in a straight line passing through S and L. The distance from the observer O to the lens L is denoted by D_L and the distance from the source S to the lens L is denoted by D_{LS} as in Fig. 8.12.

By symmetry, the conclusion drawn by Einstein is that in the above situation, the observer would perceive a 'luminous circle' having a specific *angular radius* β which by using a bit of geometry (see Appendix IX) is given by

$$\boxed{\beta = \sqrt{\frac{4GM}{c^2} \frac{D_{LS}}{D_L(D_{LS} + D_L)}}}, \qquad (8.11)$$

where M is the mass of the lens. When observed this luminous circle is now called an *Einstein ring*.

But Einstein did not believe this luminous circle would ever be observed. "Of course, there is no hope of observing this phenomenon directly... the light coming from the luminous circle cannot be distinguished by an observer as geometrically different from that coming from the star B, but simply will manifest itself as increased apparent brightness of B."

However, this increase in apparent brightness is observed in a 'microlensing' event when one star is seen to eclipse another star as viewed from Earth as in Fig. 8.13. This is a very rare event and must be done via sky surveys.

Gravitational microlensing has been used to detect extrasolar planets as the latter will show up as an additional blip in the light curve of the microlensed star when the

---

[22] Needless to say, this confirmation of Einstein's theory caused an international sensation and represents one of the greatest scientific achievements ever.

[23] For all the details see: D. Kennefick, Not Only Because of Theory: Dyson, Eddington and the Competing Myths of the 1919 Eclipse Expedition, *Einstein and the Changing Worldviews of Physics*, Springer 2011, 201–232.

[24] Lens-like action of a star by the deviation of light in the gravitational field, *Science*, 1936, Vol. 84 (1936), 505–506.

Gravitational Redshift

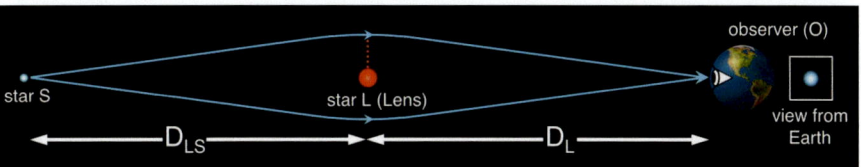

Fig. 8.12 The lensing of star S by the gravitational field of star L as viewed by the observer at O. (Courtesy Jonathan Park)

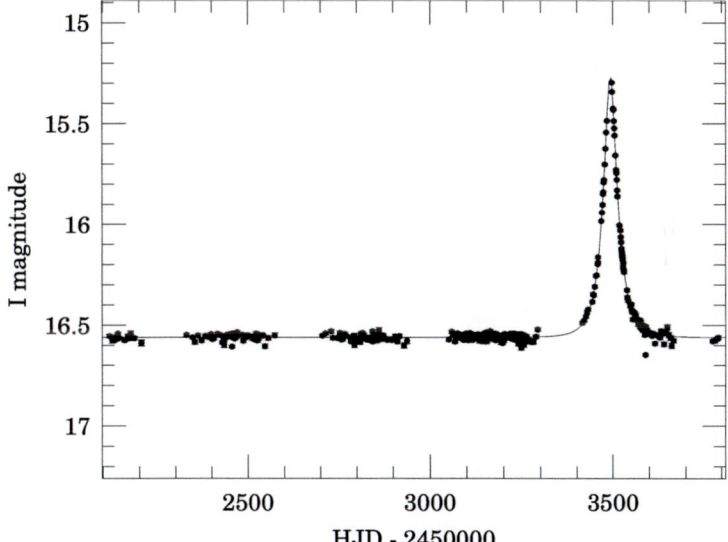

Fig. 8.13 Example light curve of the gravitational microlensing event – OGLE-2005-BLG-006. (Courtesy OGLE)

planet passes into a specific region[25] where the light of the orbited star becomes temporarily enhanced as in Fig. 8.14.

In the case of the observer being slightly off the center line through star S and lens L, then the observer will see star S as two point-like light sources, which are gravitationally deflected from the true geometrical position of S depicted in Fig. 8.15.

Again, depending on the specific geometry of the source, lens, and observer, further multiple images have been observed with the advantage that the distant object can appear brighter than otherwise as in Fig. 8.16.

---

[25] Known as the 'caustic region', a term taken from optics.

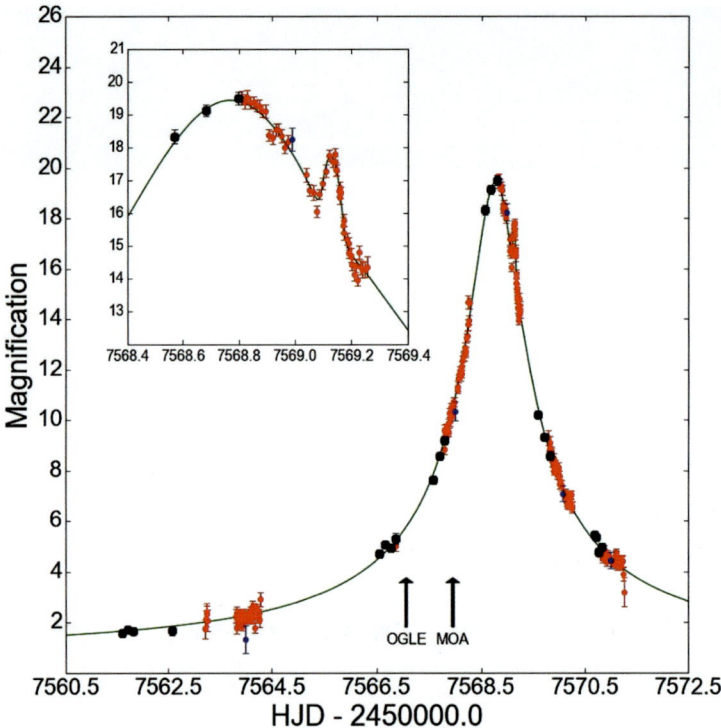

Fig. 8.14 The microlensing event of the star OGLE 2016–BLG–1195Lb. (From: I.A. Bond *et al.*, The lowest mass ratio planetary microlens: OGLE 2016–BLG–1195Lb, *Mon. Not. Roy. Astro. Soc.*, Oxford University Press, 2430–2440, 2017.) The main figure plots the data over a 12-day period where the extra-solar planetary deviation can be seen in relation to the data on other nights. The inset shows a close up of the deviation. Data from MOA in red and blue, OGLE in black. (Courtesy *Mon. Not. Roy. Astro. Soc.*, Oxford University Press, Ian Bond)

Perhaps the most striking one is known as the Einstein Cross (Fig. 8.17 (L)) which is a lensed quasar (Q2237 + 0305) residing at a distance of ~8 billion light-years, by a much closer 15th magnitude spiral galaxy (center image) at redshift $z = 0.0394$. (Reader, convert this to light-years). A faint fifth image of the quasar is also suspected to be at the position of the central galaxy. Further multiple lensed objects are in Appendix X. When the intervening lens is a galaxy or cluster of galaxies, then there is sufficient mass in Eq. (8.11) to produce an observable ring as in Fig. 8.17 (R).

Such lensed images of distant objects can determine the amount of dark matter around the lensing object. As with the Coma Cluster of Chap. 5, there is dark matter surrounding galaxies and clusters of galaxies, and the gravitational lensing effect helps in determining how much matter must be present in order to produce the lensing affects. Since the visible matter is insufficient, the rest must be dark matter.

Gravitational Redshift

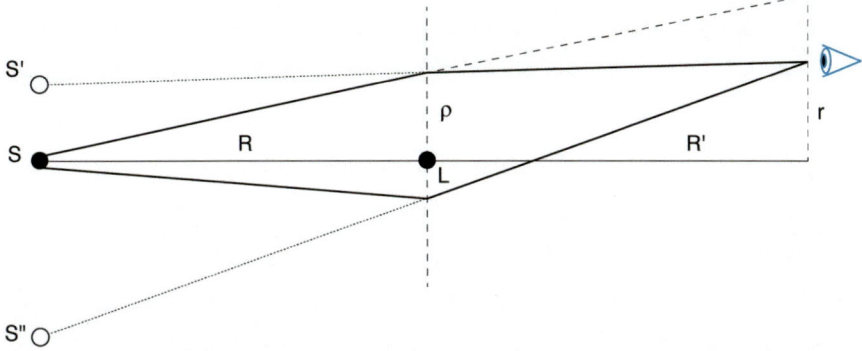

Fig. 8.15 If the observer is slightly off the center line, the result is two distant images of the source star. (Courtesy Katy Metcalf)

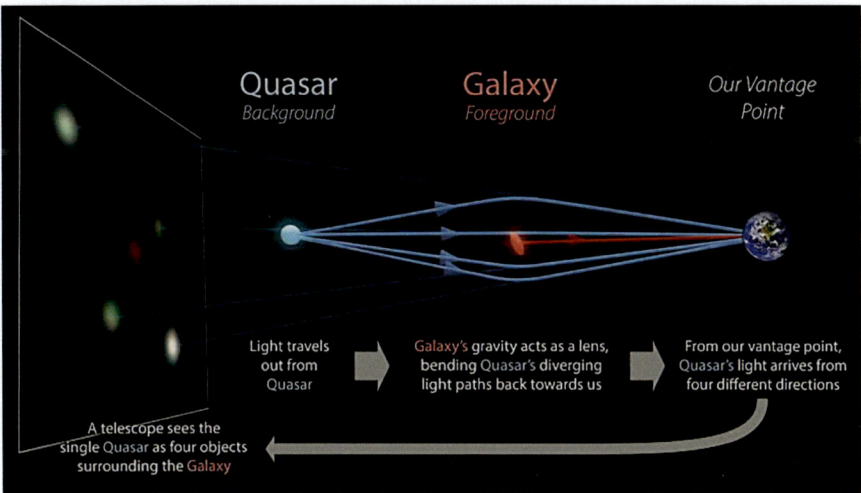

Fig. 8.16 Illustration how the light from a distant galaxy, star, or in this case a quasar can be bent by the gravitational field of an intervening galaxy. This has the effect of making the distant object brighter than it would look otherwise and can appear as multiple images as in this case). (Courtesy R. Hurt (IPAC/ Caltech)/The GraL Collaboration/ ESA)

Fig. 8.17 (L) The Einstein Cross is the four images of the gravitationally lensed quasar Q237 + 030 by the galaxy known as the Huchra Lens after astronomer John Huchra a leading member of the team who discovered it. (R) The Einstein ring LRG 3–757 showing a galaxy in the foreground (center) distorting a much more distant galaxy into a ring. (Courtesy (L) NASA/STScI (R) ESA/Hubble & NASA)

Chapter 9
Black Holes

Black holes are arguably the most interesting and mysterious objects in the Universe. They are regions of spacetime where the gravity is so intense that nothing, not even light or any other form of electromagnetic radiation can escape from within. They are found at the heart of nearly all galaxies and black holes can also collide and merge into an even larger ones. They feed off nearby gas, dust, and stars with material spiraling in towards the black hole forming an accretion disk whose inner region heats up due to friction to millions of degrees and in the process giving off powerful X-rays.

The first detected black hole was Cygnus X-1 in 1964, an intense pulsating X-ray source in the constellation of Cygnus. The 'mass' of the black hole is inferred by the gravitational effect it has on the nearby stars or if this is not possible by an empirical equation discussed below. The black holes found at the heart of most galaxies are called *supermassive black holes* (SMBH) with masses of hundreds of thousands to billions of solar masses. As well, there are much smaller ones possessing a few tens of solar mass that are called *stellar mass black holes*, and in between these two are *intermediate mass black holes* having a mass of hundreds to thousands of solar masses.

After the publication of Einstein's General Theory of Relativity in 1915, it was a year later that the German physicist/astronomer Karl Schwarzschild was able to deduce the existence of a spherical object from which nothing, not even light could escape. The boundary of this spherical object is known as the *event horizon*. The object itself is known as a *Schwarzschild black hole* and has mass, but is non-rotating and has no electric change. The latter assumption is very likely correct, but the former assumption is not, as all observed stars rotate and so it is presumed that the resulting black hole does as well. At the center of the black hole is a *singularity* where spacetime has infinite density and curvature, and theoretically at least, where all the matter of the black hole resides.[1] But the presence of the singularity indicates that current theory is incomplete and needs to be modified or extended. And to this day scientists await a further theory of quantum gravity for a more complete description of what is taking place beyond the event horizon.

Schwarzschild Radius

If the remnant core of a collapsing star has a final mass of about 2 to 3 solar masses, it will continue to collapse due to gravity and a black hole forms. Schwarzschild was able to derive the *Schwarzschild radius*[2] of the spherical black hole, that is, the radius (in meters) of the event horizon given by

$$\boxed{R_S = \frac{2GM_{BH}}{c^2}}, \tag{9.1}$$

where G is the gravitational constant, M_{BH} the mass of the black hole in kg, and c the velocity of light. A more user-friendly version can be given in terms of solar masses, that is,

$$R_S = 3\frac{M_{BH}}{M_\odot}, \tag{9.2}$$

where R_S is in kilometers. Hence, if say, a star remnant were 3 solar masses, i.e., $M_{BH} = 3M_\odot$, the resulting black hole would have a Schwarzschild radius of 9 km, whereas our Sun would have a Schwarzschild of 3 km. Both of the above equations show that the radius is proportional to the mass.

Let us just verify that Eq. (9.2) is essentially the same as Eq. (9.1). For the former we have

---

[1] An example of a mathematical singularity is given by the function $y = 1/x^2$. Then, as $x \to 0$, $y \to \infty$, which makes $x = 0$ a singularity. Here 'infinity' is just the concept of y becoming arbitrarily large as x becomes arbitrarily small, and not a quantity that can ever be attained.

[2] Strictly speaking, the Schwarzschild radius is only for a nonrotating, uncharged black hole. For other types of black holes that are rotating or have charge, other metrics must be employed and are discussed in Chap. 10.

$$R_S = \frac{2GM_{BH}}{c^2} = \frac{2 \times (6.67 \times 10^{-11} \text{ m}^3/\text{kg} \cdot \text{s}^2) M_{BH}}{9 \times 10^{16} \text{ m}^2/\text{s}^2}$$
$$\approx \left(1.5 \times \frac{10^{-27} \text{ m}}{\text{kg}}\right) M_{BH} = \left(1.5 \times \frac{10^{-30} \text{ km}}{\text{kg}}\right) M_{BH}.$$

For Eq. (9.2), we have

$$R_S = 3\frac{M_{BH}}{M_\odot} \approx \frac{(3 \times M_{BH})}{2 \times 10^{30} \text{ kg}} \text{ km} = \left(1.5 \times \frac{10^{-30} \text{ km}}{\text{kg}}\right) M_{BH},$$

and the two formulations for R_S are virtually the same, that is,

$$\frac{2G}{c^2} \approx \frac{3}{M_\odot}.$$

Exercise Compute the Schwarzschild radius if the Moon were to shrink to a black hole. *Ans.* 0.1 mm.

Exercise Compute the mass of a black hole whose radius is the Planck length, that is $l_P = 1.616\,255 \times 10^{-35}$ m. *Ans.* $\sim 10^{-5}$ gm.

From Eq. (9.1) we have the surface area of the black hole

$$A = 4\pi R_S^2 = 4\pi \left(\frac{2GM_{BH}}{c^2}\right)^2,$$

or

$$\boxed{A = \frac{16\pi G^2}{c^4} M_{BH}^2}. \tag{9.3}$$

Exercise† Compute the Schwarzschild radius of Sgr A* from the data in Chap. 5. *Ans.* $\approx 1.226 \times 10^7$ km, or about 12 million kilometers.

It was not until 1963 that New Zealand mathematician Roy Kerr found a solution to Einstein's relativity equations describing a rotating black hole (a *Kerr black hole*) and Cygnus X-1 became the first black hole candidate a year later. Although rotation is the likely feature of all black holes and thus their event horizon becomes an oblate spheroid, the Schwarzschild radius is still used as a convenient and useful metric to classify the size of black holes.

Fig. 9.1 The image of the Kerr rotating black hole at the core of the supergiant elliptical galaxy M87. Image released in 2019. (Courtesy, Event Horizon Telescope)

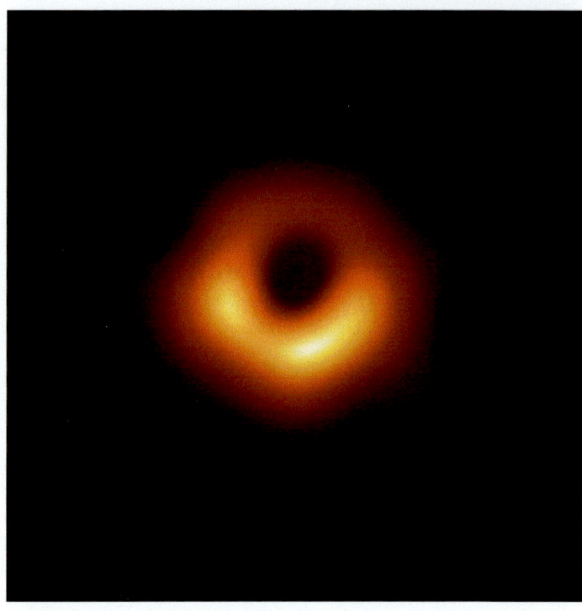

M87's Black Hole

The stunning Fig. 9.1 is the first image of a black hole taken in 2019 by the Event Horizon Telescope, which is an array of radio telescopes situated across the globe, and all synchronized by atomic clocks. This supermassive Kerr rotating black hole some 38 billion kilometers in diameter lies at the heart of the supergiant elliptical galaxy Messier 87 and is surrounded by a rapidly rotating ring of glowing gas and dust.

Example Let us calculate some features of this most interesting galaxy.

(a) Distance via TRGB method: The I band luminosity has been measured at $m = 27.10$.[3] Therefore,

$$(m - M)_I = 27.10 + 4.05 = 31.15.$$

The only added detail is converting $(m - M)_I$ to our usual visual band distance modulus. Since infrared is less affected by the dimming of gas and dust than visual light, we require a minor adjustment to obtain the visual distance modulus: $(m - M)_0 = (m - M)_I - 0.03$, to that

---

[3] S. Bird et al., The inner halo of M87: A first direct view of the red giant population, *A&A*, 524, A71 (2010), 9pp. The authors average their TRGB distance with three others to obtain a distance of 16.4 Mpc = 53.5 million light-years.

M87's Black Hole

$$(m - M)_0 = 31.12.$$

Then the distance is given by

$$d = 10^{(31.12+5)/5} = 10^{7.22} \approx 16.6 \text{ Mpc} \approx 54 \text{ Mly}$$

(b) Distance from the redshift and Hubble's Law: Redshift measured at 0.00428. Then the recessional velocity is given by

$$v_r = cz = 299{,}792.5 \text{ km/s} \times 0.00428 = 1283 \text{ km/s}.$$

The Hubble distance is

$$D = \frac{v_r}{H_0} = \frac{1283}{73} = 17.575 \text{ Mpc}$$

≈ 57 million light-years which is very close to our TRGB determination in part (a)

(c) Angular Schwarzschild diameter is ~ 15.6 μas. By Eq. (1.3) of Chap. 1,

$$\text{angular size} = \theta = 15.6 \text{ } \mu \text{ as} \times (4.848 \times 10^{-12} \text{ rad}/\mu\text{as}) = 75.63 \times 10^{-12} \text{ rad}.$$

(d) Diameter of black hole we can derive from Eq. (6.12)

$$\text{diameter} = \text{distance} \times \theta.$$

In this instance, let us use the average quoted distance to M87 of 53.5 million light-years. This equates to[4]

$$5.35 \times 10^7 \text{ ly} \times (9.462 \times 10^{12} \text{km/ly}) = 5.06 \times 10^{20} \text{ km}.$$

Therefore,

$$\text{diameter} = (5.06 \times 10^{20} \text{ km})(75.63 \times 10^{-12})$$
$$\approx 38 \times 10^9 \text{ km}$$

This gives a radius of approximately 19 billion km.

---

[4] Recall this is derived from the number of seconds in 1 year: 3.156×10^7 s/yr, giving $3.156 \times 10^7 \frac{s}{yr} \times 299{,}792.458 \frac{km}{s} = 9.462 \times 10^{12} \frac{km}{yr} = 1$ light-year.

Exercise The bright ring around the black hole at the center of the Milky Way, Sagittarius A*, is at a distance from Earth of ~26,670 light-years and has an angular diameter of 51.8 μas. Calculate its diameter in kilometers. *Ans.* ~63.4 million km.

Central Velocity Dispersion

Even though a black hole is not visible, it does have a gravitational effect on nearby stars. If some of those stars are visible then we can compute the mass of the black hole via Kepler's 3rd Law as was done in Chap. 5 for the black hole at the heart of the Milky Way, Sgr A*. However, in other galaxies the closest stars are often not visible even with the most powerful telescopes and so bright stars much further out in the galaxy must be used. The velocity of these stars is measured by their redshift having adjusted for the velocity of the galaxy itself (Fig. 9.2) due to the Hubble flow.

What is then calculated is the dispersion of the velocity of a group of stars from their average velocity. This is called the *central velocity dispersion* and is the standard deviation, σ, discussed in Chap. 1. To calculate the mass M_{BH} of the black hole at the center of a galaxy in this fashion, there are empirically derived

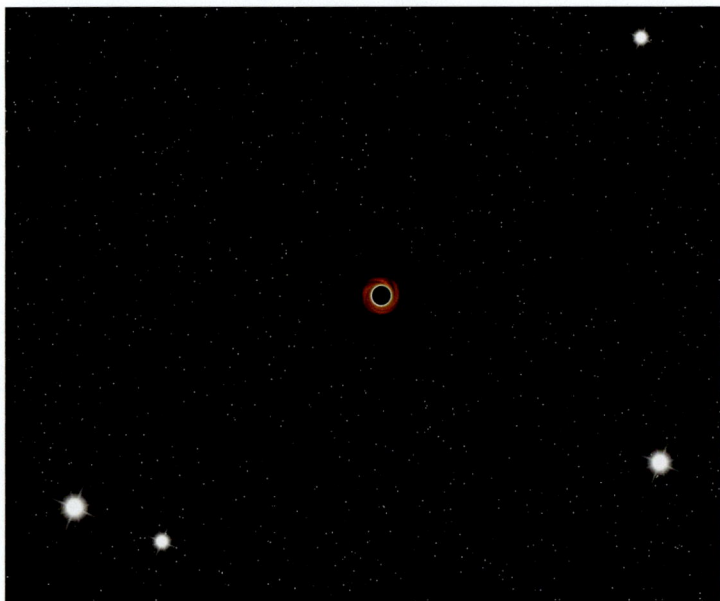

Fig. 9.2 To calculate the mass of a black hole at the core of a distant galaxy, stars far from the central black hole are required as the stars closest to the black hole are not visible. (Courtesy Jonathan Park)

formulae based on the central velocity dispersion which we denote by σ_* to indicate its stellar significance. One such version is[5]

$$\boxed{\log_{10}\left(\frac{M_{BH}}{M_\odot}\right) = 4.24 \times \log_{10}\left(\frac{\sigma_*}{200 \text{ km/s}}\right) + 8.36}. \quad (9.4)$$

Observe that this is just the form of a straight line: $y = ax + b$, where:

$$y = \log_{10}\left(\frac{M_{BH}}{M_\odot}\right);$$

$$x = \log_{10}\left(\frac{\sigma_*}{200 \text{ km/s}}\right);$$

$$a = 4.24, b = 8.36;$$

and M_{BH} is the mass of the black hole, $M_\odot$ is the solar mass, and σ_* is the central velocity dispersion measured in km/s. Note that as the black hole increases in mass, so must the central velocity dispersion σ_*.

Going back to the galaxy M87, it has been found to have a central velocity dispersion of $\sigma_* = 440$ km/s. Putting this value into our Eq. (9.4)

$$\log\left(\frac{M_{BH}}{M_\odot}\right) = 4.24 \times \log_{10}\left(\frac{440 \text{ km/s}}{200 \text{ km/s}}\right) + 8.36$$
$$= (4.24 \times 0.342) + 8.36 = 9.81.$$

Therefore, the mass of the central black hole in M87 is

$$M_{BH} = 10^{9.81} \approx 6.5 \text{ billion solar masses}.$$

It should be mentioned that there is a reasonable amount of scatter in the data and some difficulty in pinning down the coefficients a and b in deriving formulas such as in footnote 5.

Exercise Determine the Schwarzschild radius of the central black hole of M87 from Eq. (9.1) with the value of the gravitational constant given by $G = (6.67 \times 10^{-11})$ m^3/kg·s^2. Ans. ~19.2 billion km.

---

[5] Adapted by the author for illustration purposes from the least squares regression formula:

$$\log_{10}\left(\frac{M_{BH}}{M_\odot}\right) = 4.38 \times \log_{10}\left(\frac{\sigma_*}{200 \text{ km/s}}\right) + 8.49$$

in Kormendy and Ho, Coevolution (Or Not) of Supermassive Black Holes and Host Galaxies, Ann. Rev. A&A, 51 (2013), 511–653.

Exercise The galaxy NGC 6861 has a velocity dispersion value of $\sigma_* = 414$ km/s. Using the formula of footnote 5 determine the mass of its central black hole. *Ans.* ~7.5 billion solar masses.

Exercise The galaxy NGC 6868 has a velocity dispersion value of $\sigma_* = 250$ km/s. Using the formula of footnote 5 determine the mass of its central black hole. *Ans.* ~820 million solar masses.

Black Hole Density

The density of an object of mass M and radius R is its mass divided by its volume V where

$$V = \frac{4}{3}\pi R^3.$$

So let us determine the density of the black hole at the center of M87, although we must appreciate that this does not in any way represent the physical reality of what is going on inside the black hole.

Firstly, the Schwarzschild radius is given by Eq. (9.2)

$$R_S = 3 \times (6.5 \times 10^9) = 1.95 \times 10^{10} \approx 2 \times 10^{10} \text{ km}.$$

The mass of the black hole is 6.5 billion times the mass of the Sun (2×10^{30} kg), that is,

$$\begin{aligned} M &= (6.5 \times 10^9) \times (2 \times 10^{30} \text{ kg}) \\ &= 1.3 \times 10^{40} \text{ kg}. \end{aligned}$$

The volume is (using the Schwarzschild radius and assuming a spherical shape)

$$V = \frac{4}{3}\pi \times (2 \times 10^{10} \text{km})^3 = 3.4 \times 10^{31} \text{km}^3,$$

and dividing the mass by the volume we find that the density is

$$\rho = \frac{1.3 \times 10^{40} \text{ kg}}{3.4 \times 10^{31} \text{ km}^3} \approx 0.4 \times 10^9 \text{ kg/km}^3.$$

This is a very sparce density so let us convert this to gm/cm^3 which is a bit more meaningful, giving

$$\rho \approx 0.0004 \, \text{gm/cm}^3,$$

where air at sea level is about 0.001225 gm/cm³. So, our black hole is much less dense than air!

This dispels the notion that a black hole must be exceedingly dense. Actually, the larger the mass of the black hole the less dense it becomes. To see this, note that by Eqs. (9.1) or (9.2) that the Schwarzschild radius is directly proportional to the mass of the black hole so that doubling the mass is the same as doubling the radius. On the other hand, the volume of the black hole increases with the cube of the radius, that is, doubling the radius increases the volume by 8 times. Therefore, if we double the radius then as the density is mass divided by volume, the mass is doubled but the volume is increased by a factor of 8. This means that as the black hole increases in size, its density decreases.

Exercise Show that in general the density of a black hole is given by the equation

$$\rho = \frac{3c^6}{32\pi G^3 M^2}.$$

Exercise Determine the density of the Universe given a total mass of 3.4×10^{54} kg and a radius of 46.5 billion light-years. *Ans.* 9.5×10^{-30} gm/cm³.

On the other hand, a typical neutron star *is* exceedingly dense. A typical neutron star has a radius of 10 km and mass approximately at the Chandrasekar limit: $1.4 \, M_\odot = 2.784 \times 10^{30}$ kg, discussed in Chap. 8. The volume is

$$V = \frac{4}{3}\pi \times (10 \text{ km})^3 = 4189 \text{ km}^3 = 4.189 \times 10^{12} \text{ m}^3,$$

implying that the density is

$$\rho = \frac{2.784 \times 10^{30} \text{ kg}}{4.189 \times 10^{12} \text{ m}^3} = 6.65 \times 10^{17} \text{ kg/m}^3 = 6.65 \times 10^{14} \frac{\text{gm}}{\text{cm}^3}.$$

Comparing this with the mass of the Sun which we calculated in Chap. 1 to be $\rho = 1.41$ gm/cm³ we see that a typical neutron star is 470 trillion times denser than the Sun.

Schwarzschild Precession for Black Holes

In Eq. (8.5), we had for the Schwarzschild precession

$$\delta\phi = \frac{6\pi GM}{c^2 a \left(1 - e^2\right)},$$

and if we take the mass M to be that of a black hole, then we recognize the quantity

$$\frac{2GM}{c^2} = R_S,$$

as the Schwarzschild radius. This means that we have a further expression for the Schwarzschild precession, namely

$$\boxed{\delta\phi = \frac{3\pi R_S}{a(1-e^2)}} \tag{9.5}$$

in radians per orbit.

Exercise
(a) In Chap. 8 we calculated the Schwarzschild precession of the star S2 that orbits the black hole Sgr A* at the center of the Milky Way. Its Schwarzschild radius is $R_S = 12.26 \times 10^6$ km from Exercise†. Using Eq. (9.5) with semi-major axis $a = 15.35 \times 10^{10}$ km and $e = 0.885$, you should obtain the same result of ≈ 12 arcmin/orbit for its Schwarzschild precession.
(b) Compute the Schwarzschild precession in arcmin per orbit for the star S62 orbiting the black hole Sgr A* with semi-major axis $a = 11.075 \times 10^{10}$ km and eccentricity $e = 0.9760$. Ans. 75.6 arcmin/orbit.
(c) Calculate the Schwarzschild precession for the star S4711 that has a semi-major axis $a = 619.2$ AU and eccentricity $e = 0.768$. Ans. ~10.5 arcmin/orbit.

Hawking Radiation/Temperature

A most interesting thing happens near the event horizon of a black hole where Quantum Mechanics meets Cosmology. From quantum field considerations Stephen Hawking was able to deduce that black holes emit radiation now known as *Hawking radiation*. This is the case because 'empty space' is actually roiling with activity from what are called *virtual particles* which are not particles at all in the conventional sense, but mathematical constructs used to explain the intermediate steps in real particle interactions.[6] They are created spontaneously often as a pair of a particle and anti-particle which very quickly annihilate one another.[7]

However, at the event horizon of a black hole, a virtual particle pair can become separated due to the intense gravity of the black hole and one particle will fall into the black hole but the other escape. The escaping particle is emitted as thermal

---

[6] Yes, this sounds strange but so is the nature of Quantum Mechanics!
[7] The separation time is determined by the Heisenberg Uncertainty Principle and are typically on the order of 10^{-21} to 10^{-24} seconds or less.

Hawking Radiation/Temperature

Fig. 9.3 Black holes not only devour matter in its vicinity but also are the source of radiation. (Courtesy NASA)

radiation (*Hawking radiation*) at a particular temperature that depends on the mass of the black hole (see Fig. 9.3). Specifically, the black body Hawking radiation has a temperature given by[8]

$$\boxed{T_H = \left(\frac{\hbar c^3}{8\pi k_B G}\right) \cdot \frac{1}{M_{BH}}}, \tag{9.6}$$

involving the usual constants that we have seen before. Note the interesting feature of the *Hawking temperature* is that it is *inversely proportional* to the mass of the black hole so that smaller black holes radiate at a higher temperature. Furthermore, this emission of radiation leads to the mass of the black hole gradually diminishing over eons of time but with rising temperature.

For example, taking a black hole with a mass of $M_{BH} = 10 M_\odot = 1.9885 \times 10^{31}$ kg. Then we have for the numerator of Eq. (9.6)

---

[8] Or equivalently, $T_H = \left(\frac{hc^3}{16\pi^2 k_B G}\right) \cdot \frac{1}{M_{BH}}$.

$$\hbar c^3 = \left(1.0546 \times 10^{-34} \text{ m}^2 \cdot \text{kg/s}\right) \times \left(2.9979 \times 10^8 \text{ m/s}\right)^3$$
$$= 28.414 \times 10^{-10} \text{ m}^5 \cdot \text{kg/s}^4.$$
$$8\pi k_B G = (8 \times 3.14159) \times \left(1.3806 \times 10^{-23} \text{ kg} \cdot \text{m}^2/\text{s}^2 \cdot \text{K}\right)$$
$$\times \left(6.6743 \times 10^{-11} \text{ m}^3/\text{kg} \cdot \text{s}^2\right) = 231.59 \times 10^{-34} \text{ m}^5/\text{s}^4 \cdot \text{K}.$$

Dividing the numerator by denominator,

$$\frac{\hbar c^3}{8\pi k_B G} = \frac{28.414 \times 10^{-10} \text{m}^5 \cdot \text{kg/s}^4}{2.3159 \times 10^{-32} \text{ m}^5/\text{s}^4 \cdot \text{K}} = 12.27 \times 10^{22} \text{ kg} \cdot \text{K}.$$

Finally, we insert the mass of the black hole to obtain

$$T_H = \left(\frac{\hbar c^3}{8\pi k_B G}\right) \cdot \frac{1}{M_{BH}} = \frac{12.27 \times 10^{22} \text{ kg} \cdot \text{K}}{1.989 \times 10^{31} \text{ kg}} \times 10^{22} = 6.17 \times 10^{-9} \text{ K}.$$

A black hole that is $30 M_\odot$ will radiate at 1/3 of this value.[9]

Black Hole Luminosity

Considering a black hole as a blackbody, we can use this temperature to calculate the luminosity of a black hole from Eq. (3.7)

$$L = 4\pi R_S^2 \cdot \sigma T_H^4 = 4\pi \left(\frac{2GM_{BH}}{c^2}\right)^2 \left(\frac{2\pi^5 k_B^4}{15 c^2 h^3}\right) \left(\frac{hc^3}{16\pi^2 k_B G M_{BH}}\right)^4$$
$$= \left(\frac{hc^6}{30720 \pi^2 G^2}\right) \cdot \frac{1}{M_{BH}^2} = \left(\frac{\hbar c^6}{15360 \pi G^2}\right) \cdot \frac{1}{M_{BH}^2}.$$

We note however that this formula does not take into account any quantum mechanical effects in the context of gravity and awaits a complete theory of quantum gravity. So, it is appropriate to call L in this case the *nominal luminosity*.

Exercise Compute the luminosity of a black hole of mass $M_{BH} = 10 M_\odot$ from the preceding considerations. *Ans.* 9×10^{-31} W.

---

[9] The reader can find a Hawking radiation calculator at the website of Victor Toth: https://www.vttoth.com/CMS/physics-notes/311-hawking-radiation-calculator

Black Hole Evaporation

Since a black hole radiates and as a consequence is gradually losing mass, then its mass becomes a function of time and therefore a time exists when it has completely evaporated. This time is given by

$$\boxed{t_{ev} = \frac{5120\pi G^2 M_0^3}{hc^4}}, \tag{9.7}$$

where here M_0 is the initial mass of the black hole. Hence, we can put in a black hole mass and see how long it will take to evaporate into nothing.

For example, if the black hole mass is one solar mass, then we obtain a time of $\sim 10^{67}$ years which vastly exceeds the age of the Universe. On the other hand, for a very small black hole, say $M_0 = 10^{11}$ kg, then we obtain

$$t_{ev} = \frac{5120 \times 3.14159 \times (6.6743 \times 10^{-11} \text{ m}^3/\text{kg} \cdot \text{s}^2)^2 \times (10^{11} \text{ kg})^3}{(1.0547 \times 10^{-34} \text{ m}^2 \cdot \text{kg/s}) \times (2.998 \times 10^8 \text{ m/s})^4} \approx 8.4 \times 10^{16} \text{ s}.$$

Converting this to more meaningful years with 1 year $= 3.156 \times 10^7 s$, yields $t_{ev} \approx 2.7$ billion years, well within the age of the Universe.

In simple terms the evaporation of black holes leads to a paradox since a black hole contains information and once it has evaporated all that information is lost. This is the *black hole information paradox* and its resolution is an active area of discussion and is a part of the ongoing research into quantum gravity.

Exercise If a black hole formed at the time of the Big Bang 13.8 billion years ago, what would its mass have been if it has just evaporated? *Ans.* $\sim 1.7 \times 10^{11}$ kg.

Black Hole Entropy

Entropy is an interesting property that has many interpretations in a diverse range of fields. Let us simply consider it as a measure of the disorder of a system and phrase the *Second Law of Thermodynamics* such that: *the entropy of a closed system cannot decrease*. Because of the nature of a black hole, it was once thought that it would have no entropy. However, black holes can pull in hot gas from its surrounding neighborhood and this gas will be highly disordered, thus having a high level of entropy. If all that entropy vanished once inside the black hole, this would violate the 2nd Law. The apparent contradiction was resolved by astrophysicists Jacob Beckenstein and Stephen Hawking who demonstrated that a black hole does indeed have entropy, denoted by S_{BH} and given by

$$\boxed{S_{BH} = \frac{k_B c^3}{4G\hbar} A = \frac{k_B}{4l_P^2} A}, \tag{9.8}$$

where $A = 4\pi R_S^2$ is the surface area of the event horizon and the other constants such as the Boltzmann constant k_B etc. are again familiar. As well, we are using the formula for Planck Length, $l_P^2 = \hbar G/c^3$.

This result is particularly interesting as it says that all the information about the entropy of a black hole depends only on the surface area of its event horizon. From this idea, the physicist Gerard 't Hooft and Leonard Susskind developed the *holographic principle* whereby the physics that occurs in a region of space can be described by the information contained in the boundary surface of the region. For those readers who know about the Poisson Formula (for harmonic functions) that gives the (steady-state) temperature at any point inside a spherical region in terms of the temperature values on the boundary of the sphere, this will be a familiar concept.

Exercise Express the black hole entropy in terms of the mass M of the black hole. Ans. $S_{BH} = \frac{4\pi k_B G M^2}{\hbar c}$.

Chapter 10
The Universe

Einstein Field Equations (EFE)

After developing his Special Theory of Relativity (1905), which does not include gravity in its considerations, Einstein spent the next ten years developing an all-encompassing General Theory of Relativity that does include gravity and essentially describes, in a series of 10 equations, how matter, energy, and space, in the Universe on the scale of planets, stars, and galaxies, behave. Two basic assumptions are made about the Universe itself on the largest scales, namely that it is *homogeneous*, so that roughly it is the same everywhere regarding its mass and energy, and it should also be *isotropic* in that it should look basically the same in all directions.[1] For such a Universe, the EFE can be summarized in the beautiful expression

$$\boxed{G_{uv} = \frac{8\pi G}{c^4} T_{uv}} \tag{10.1}$$

where G_{uv} (the *Einstein tensor*) encodes information regarding the curvature of spacetime and T_{uv} (*energy-momentum tensor*) describes the distribution of matter and energy. The subscripts u, v run through the indices 0, 1, 2, 3 where the first index represents time and the other three represent the three spatial dimensions. The EFE describe the curvature of spacetime due to the presence of mass and energy.

While the details of the theory are far beyond the scope of this book, we can detect some known constants, c the velocity of light, π which arises everywhere in Physics and Astronomy, and the gravitational constant G. It is from these field equations that the bending of light by the Sun could be calculated, sending Eddington and Dyson to remote parts of the world to verify it for themselves, and for the world. The field equations also give us gravitational time dilation, gravitational redshift, gravitational

---

[1] A good example that the Universe is isotropic is the Cosmic Background Radiation which is remarkably constant across its entire expanse.

lensing as well as black holes and the Schwarzschild radius, that have been discussed earlier.

Two years later in 1917 Einstein added an additional term called the *cosmological constant* as it was at the time believed that the Universe was static over time, neither expanding nor contracting. The effect of adding Λ into the equation was that of an outward pressure in space in order to counterbalance the gravitational effects of the accumulated mass in the Universe, which would otherwise tend to cause the Universe to contract in on itself. In order to offset this, the field equations were modified to:[2]

$$\boxed{G_{uv} + \Lambda g_{uv} = \frac{8\pi G}{c^4} T_{uv}}. \tag{10.2}$$

However, the cosmological constant was later discarded by Einstein following the discovery of the expansion of the Universe as revealed by the observational work of Edwin Hubble, Milton Humason, Vesto Slipher, and Henrietta Leavitt.[3] It is worth noting that the expansion of the Universe was an idea originally considered by the Russian physicist Alexander Friedmann, and later developed by the Belgian priest Georges Lemaître in 1927, who based his theoretical work on the recent redshift observations. The cosmological constant experienced a revival in the 1990s when it was discovered that this expansion of the Universe was accelerating.

This new found acceleration of the Universe has been linked with the roiling vacuum of space. In Quantum Mechanics, no space can be completely empty and there is a constant ferment of virtual particles as mentioned in the previous chapter, that come into existence and then annihilate each other just as quickly as they appeared. The activity gives rise to what is known as a *vacuum energy* that permeates all of space.

On the other hand, there is a major problem with associating vacuum energy as a main component of the cosmological constant Λ. When physicists computed the value of this vacuum energy from quantum theory, it was found to be astronomically large, so much so that if true, stars would be too distant for their light to even reach us. This vast discrepancy presents a big dilemma for Cosmology, and, in fact, it is known as the *cosmological constant problem*. In order to address this issue, the concept of *dark energy* was proposed, a mysterious form of energy thought to be responsible for the Universe's accelerated expansion. Many theories have attempted to reconcile dark energy with the vacuum energy and the cosmological constant Λ, but without any conclusive resolution. Despite this, it is well established that the

---

[2] A cosmological model based on Einstein's Field Equations was presented by Willem de Sitter (1917) based on the assumption that the density of the Universe was so sparce as to be essentially devoid of mass, but did retain a positive cosmological constant. This is discussed further in the sequel.

[3] "... the redshift of distant nebulae has smashed my old construction [a static closed Universe] like a hammer blow."

value of Λ is nonnegative and extremely small, and in accordance with the observed expansion rate of the Universe.[4]

Geometry of Space

Euclid's Fifth Postulate of plane geometry states that given any straight line and a point not on the line, then at most, one straight line can be drawn through the point that is parallel to the given line. An equivalent statement is that the sum of the interior angles of any triangle will always be 180 degrees. This is the geometry of a flat sheet of paper.

But if we were to draw a giant triangle between three stars, would their sum likewise be exactly 180 degrees? If we draw a triangle on a sphere, then the sum of the interior angles is greater than 180 degrees. And if we similarly draw a triangle on the surface of a saddle, then the sum of the interior angles will be less than 180 degrees.

So, what is going on with our Universe? To see what is happening, we need to go back to an equation due to Alexander Friedmann in the 1920s, which basically laid the foundations for modern Cosmology. This he derived from the Einstein Field Equations and is based on both the geometry of the Universe and the amount of matter it contains. It leads to a dynamic Universe that had hitherto been thought to be static. The key ingredient here is the time-dependent *scale factor*, $a(t)$[5] which describes how space expands or contracts.

To review, as discussed in Chap. 7, suppose that at a particular moment in time, t_0, that our Milky Way galaxy and the galaxy M83 are a distance $D(t_0)$ apart from one another. As the Universe expands over a time t, the distance between the two galaxies is now $D(t)$, which can be expressed as

$$D(t) = a(t)D(t_0). \tag{10.3}$$

Taking the derivative of this equation, we can compute the rate of change (recessional velocity) v at which the two galaxies are rushing away from one another due to the expansion of the Universe, namely

---

[4] Recent results (2024) from the Dark Energy Spectroscopic Instrument (DESI) experiment hint that dark energy could in fact be weakening. If this data holds up it would suggest that the behavior of dark energy is more complex than that given by a simple constant. This opens up the possibility of other theories, such as *quintessence*, in which Λ is replaced by a dynamic scalar field ϕ, whose energy density ρ_ϕ, varies over time.

[5] Sometimes this is denoted by $R(t)$.

$$v = \frac{dD(t)}{dt} = \dot{a}(t)D(t_0),$$

since $D(t_0)$ is a fixed value and $\dot{a}(t)$ represents the derivative of the scaling factor $a(t)$. This leads to

$$D(t) = a(t)\frac{v}{\dot{a}(t)},$$

and solving for the velocity v,

$$v = \frac{\dot{a}(t)}{a(t)}D(t).$$

Setting

$$\boxed{H_t = \frac{\dot{a}(t)}{a(t)}}, \quad (10.4)$$

yields Hubble's Law

$$v = H_t \cdot D(t).$$

Our conventional use of Hubble's Law has the 'Hubble Parameter' denoted by H_0 which denotes the present era. In the sequel we will also use H for H_t.

First Friedmann Equation

The main Friedmann equation involves the square of the Hubble Parameter and the scale factor $a(t)$ and is expressed by

$$H^2 = \frac{8\pi G}{3}\rho - \frac{kc^2}{a^2},$$

where H is the Hubble Parameter, ρ is the mass-energy density. Here k is a *curvature parameter* which can take the values: $k = 0$, $k > 0$, $k < 0$.[6] The the mass-energy density represents the total amount of matter and radiation contained in a given

---

[6] The curvature parameter k does not measure the curvature in the same sense as Gaussian curvature (i.e., $1/R^2$) used to describe geometric surfaces, as k also involves the scale factor $a(t)$.

First Friedmann Equation

volume of space. A general discussion of the notion of (Gaussian) curvature is discussed in Appendix XI.

The preceding formulation was the original version given by Friedmann, but modern versions include Einstein's cosmological constant Λ to account for the accelerating expansion of the Universe:

$$\boxed{H^2 = \frac{8\pi G}{3}\rho - \frac{kc^2}{a^2} + \frac{\Lambda c^2}{3}}. \tag{10.5}$$

In this equation, ρ includes all normal (baryonic) and dark matter, and all forms of radiation. We can write this as:

$$\rho = \rho_{baryonic\ matter} + \rho_{dark\ matter} + \rho_{radiation}$$

since we are already accounting for the density effects of dark energy in the Λ term. According to the Einstein Field Equations, the energy density associated with Λ is given by

$$\boxed{\rho_\Lambda = \frac{\Lambda c^2}{8\pi G}}. \tag{10.6}$$

It is important to note that this value is strictly constant and does not diminish with time in spite of the expansion of the Universe, unlike ordinary matter (which scales as $\rho \propto a^{-3}$), and for radiation (which scales as $\rho \propto a^{-4}$), with a being the scale factor.[7]

Thus, if we incorporate the dark energy density into the expression for ρ, namely

$$\rho = \rho_{total} = \rho_{baryonic\ matter} + \rho_{dark\ matter} + \rho_{radiation} + \rho_\Lambda,$$

then we can write Eq. (10.5) in its original form

$$\boxed{H^2 = \frac{8\pi G}{3}\rho - \frac{kc^2}{a^2}}, \tag{10.7}$$

since by Eq. (10.6) for ρ_Λ

---

[7] The reason that in the case of radiation, the energy density ρ is proportional to a^{-4} (rather than simply a^{-3}) is due to the redshift of the radiation as the Universe expands resulting in the radiation having proportionately longer wavelength, that is, $\lambda \propto a$. Since $E = \frac{hc}{\lambda}$, where λ is the wavelength, this results in, $\rho \propto a^{-3} \cdot a^{-1} = a^{-4}$.

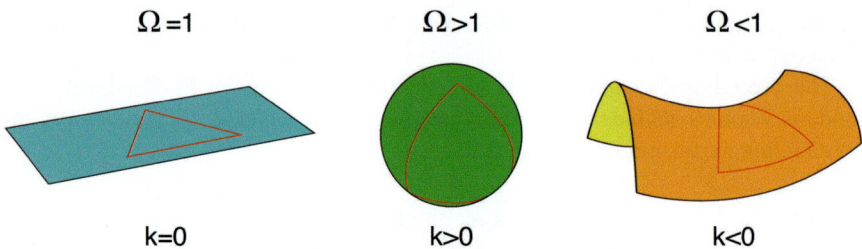

Fig. 10.1 The three potential geometries of the Universe: spherical, hyperbolic or flat. Each geometry depends on the matter-energy density of the Universe. (Courtesy Katy Metcalf)

$$\frac{8\pi G}{3}\rho_\Lambda = \frac{\Lambda c^2}{3}$$

already accounts for the Λ dark energy term in the modern Friedmann formulation, Eq. (10.5). Therefore, what is taken for the mass-energy density ρ needs to be in accordance with the form of the version of the Friedmann Equation that is being employed.

When $k = 0$: The Universe is termed *flat* and the geometry is completely Euclidean. The interior angles of giant triangles in space add up to 180°;

When $k > 0$: The Universe is positively curved and closed, analogous to the surface of a sphere and interior angles of giant triangles in space add up to greater than 180°;

When $k < 0$: The Universe is negatively curved and open, analogous the surface of a saddle and interior angles of giant triangles in space add up to less than 180°. See Fig. 10.1 for the three possible geometries.

Eras Tour

In order to solve the Eq. (10.7) let us make a few simplifying assumptions. We will assume the geometry of the Universe is flat and take $k = 0$ as this is apparently the case in view of the next section.

I. Dark Energy-Dominated Era

In examining the most recent era we assume that the contributions from both ordinary matter and dark matter (and radiation) represented by ρ are negligible so that $\rho = \rho_\Lambda$ in our present *dark energy-dominated era* from about 9.8 billion years after the Big Bang.[8] Thus Eq. (10.7) reduces to

---

[8]This is due to the decreasing density of ordinary and dark matter with increasing volume but the dark energy density remains constant over time for each unit volume of space.

$$H^2 = \frac{8\pi G}{3}\rho_\Lambda = \frac{\Lambda c^2}{3}.$$

Since we are not concerned with units here, take $c = 1$, which is a common practice so that by Eq. (10.4)

$$H = \frac{\dot{a}(t)}{a(t)} = \frac{\frac{da}{dt}}{a(t)} = \sqrt{\frac{\Lambda}{3}} = C. \qquad (10.8)$$

This is a first-order linear differential equation and separating the variables we have

$$\frac{da}{a} = C\, dt.$$

Upon integrating both sides

$$\int \frac{da}{a} = C \int dt$$

gives

$$\log a = Ct + C_1.$$

Exponentiating both sides results in

$$a(t) = C_2 e^{Ct},$$

where $C_2 = a(0) = a_0$ and $H = H_0$.[9] Therefore, we can write

$$a(t) = a_0 e^{Ct}.$$

Finally,

$$a(t) = a_0 e^{\sqrt{\frac{\Lambda}{3}}t},$$

and recalling that $H = C = \sqrt{\frac{\Lambda}{3}}$ from Eq. (10.8) we arrive at

---

[9] In this instance, we are using $t = 0$ to represent the present time, not the time of the Big Bang. We could have instead put in $t = t_0$ but we would still derive $a(t) \propto e^{H_0 t}$. It just looks cleaner this way. Indeed, a_0 is typically set to 1.

$$a(t) \propto e^{H_0 t}. \qquad (10.9)$$

This indicates that the expansion of the Universe is increasing at an exponential rate and therefore accelerating ($\ddot{a} > 0$).

In the sequel we will mention the *de Sitter Universe* where there is no mass but only dark energy and Eq. (10.9) represents the scale factor for that era.

II. Matter-Dominated Era

Preceding the present dark energy-dominated era was the *matter-dominated era* from 47,000 years after the Big Bang when the Universe cooled sufficiently as matter began to dominate to about 9.8 billion years following. In this instance, the Friedmann equation reduces to

$$H^2 = \frac{8\pi G}{3}\rho_m$$

where ρ_m is now the matter density. This density decreases with the cube of the volume, that is, $\rho_m \propto a^{-3}$, and employing this in the preceding expression[10] and with a bit more effort one can derive analogously to the preceding case that during this period

$$a(t) \propto t^{\frac{2}{3}}. \qquad (10.10)$$

III. Radiation-Dominated Era

Prior to the matter-dominated era was the *radiation-dominated era* which lasted ~47,000 years after the Big Bang. During this time, the expansion of the Universe can again be derived from the Friedmann equation

---

[10] Since $\rho_m \propto a^{-3}$, we can set $\rho_m = \frac{\rho_0}{a^3}$, implying that

$$H^2 = \left(\frac{\dot{a}}{a}\right)^2 = \frac{8\pi G}{3}\rho_m = \rho_0 \frac{8\pi G}{3}\left(\frac{1}{a^3}\right).$$

Taking the square root of both sides, gives

$$\dot{a} = \left(\rho_0 \frac{8\pi G}{3}\right)^{\frac{1}{2}} a^{-\frac{1}{2}},$$

so that separating the variables and integrating,

$$\int a^{\frac{1}{2}} da = \left(\rho_0 \frac{8\pi G}{3}\right)^{\frac{1}{2}} \int dt.$$

Exercise. Finish the proof. Note that in this instance, $t = 0$ represents the moment of the Big Bang and $a(0) = 0$, which will imply that the constant of integration equals zero. This will give Eq. 10.10.

Second Friedmann Equation

$$H^2 = \frac{8\pi G}{3}\rho_r$$

where this time ρ_r is the radiation density. This can again be related to the scale factor $a(t)$ and solving the resulting differential equation gives

$$\boxed{a(t) \propto t^{\frac{1}{2}}}. \tag{10.11}$$

We conclude that in this era $a(t)$ is growing slower than during the matter-dominated era, and the latter's expansion rate, $H(t)$, decreased due to the gravitational attraction of matter. Then, as the density of matter further decreased, the Universe transitioned to the dark energy-dominated era in which the repulsive force of dark energy began to accelerate the Universe's expansion.

Second Friedmann Equation

A second Friedmann equation involves the acceleration of the scale factor $a(t)$

$$\boxed{\frac{\ddot{a}}{a} = \frac{-4\pi G}{3}\left(\rho + \frac{3p}{c^2}\right)}, \tag{10.12}$$

where p is the pressure of the matter and radiation, in the Universe, and initially we take $\Lambda = 0$ (no dark energy). Here, ρ includes all forms of matter and radiation, that is, $\rho = \rho_m + \rho_r$, and likewise for the pressure, $p = p_m + p_r$.

In order to include dark energy, we know that the dark energy density is given by Eq. (10.6), namely, $\rho_\Lambda = \frac{\Lambda c^2}{8\pi G}$, and moreover, the pressure is a negative one given by $p_\Lambda = -c^2 \rho_\Lambda = -\frac{\Lambda c^4}{8\pi G}$.[11] Hence, we can expand Eq. (10.12) to include dark energy by writing,

$$\frac{\ddot{a}}{a} = \frac{-4\pi G}{3}\left(\rho + \rho_\Lambda + \frac{3(p + p_\Lambda)}{c^2}\right).$$

Substituting in the values for ρ_Λ and p_Λ, we get

---

[11] The c^2 is required to ensure consistency of the units which are J/m^3 for ρ_Λ and p_Λ.

$$\frac{\ddot{a}}{a} = \frac{-4\pi G}{3}\left(\rho + \frac{\Lambda c^2}{8\pi G} + \frac{3\left(p - \frac{\Lambda c^4}{8\pi G}\right)}{c^2}\right)$$

$$= \frac{-4\pi G}{3}\left(\rho + \frac{\Lambda c^2}{8\pi G} + \frac{3p}{c^2} - \frac{3\Lambda c^2}{8\pi G}\right).$$

Upon simplifying this last expression, we have the 2nd Friedmann equation that includes dark energy, namely

$$\boxed{\frac{\ddot{a}}{a} = \frac{-4\pi G}{3}\left(\rho + \frac{3p}{c^2}\right) + \frac{\Lambda c^2}{3}} \qquad (10.13)$$

where the density ρ and the pressure p are as above.

As an example, let us assume that we have a dark energy dominated Universe as appears to be the present case, so that we assume, $\rho_m + \rho_r \approx 0$, $p_m + p_r = 0$. Then by Eq. (10.13),

$$\frac{\ddot{a}}{a} = \frac{\Lambda c^2}{3},$$

or,

$$\ddot{a}(t) = a(t)\frac{\Lambda c^2}{3} > 0.$$

Therefore, the two Friedmann equations describe the evolution and expansion of the Universe.

Exercise†† Show that $\frac{\ddot{a}}{a} = \dot{H} + H^2$, where H is the Hubble Parameter.

In Appendix XII we derive the 2nd Friedmann equation from the first one Eq. (10.5) using the conservation of energy-momentum relation for a perfect fluid in an expanding Universe:

$$\boxed{\dot{\rho} + 3\frac{\dot{a}}{a}\left(\rho + \frac{p}{c^2}\right) = 0}. \qquad (10.14)$$

Critical Density

From the above three cases for the mass-energy density we can write these in terms of another deciding factor known as the *critical density*, ρ_c, which we obtain when we set the curvature constant $k = 0$ (the Euclidean case) in the Friedmann Eq. (10.7) and solve for ρ ($= \rho_{total} = \rho_m + \rho_r + \rho_\Lambda$),

$$\boxed{\rho_c = \rho = \frac{3H_0^2}{8\pi G}} \qquad (10.15)$$

where H_0 is the Hubble Parameter for the present era, and G the gravitational constant.

Here, the critical mass-energy density ρ of the Universe depends heavily on the value of Hubble Parameter H_0, which is squared. This is one reason why the value of the Hubble Parameter is such an important number.

Let us determine just what sort of density we are talking about. Assuming a value of $H_0 = 70$ km/s/Mpc, then we find that the critical density is[12]

$$\rho_c = \frac{3H_0^2}{8\pi G}$$

$$= \frac{3 \times (70 \times 10^3 \text{ m/s/Mpc})^2}{8 \times 3.14 \times 6.67 \times 10^{-11} \text{ m}^3/\text{kg} \cdot \text{s}^2}$$

$$\approx \frac{87.73 \times 10^{17} \cdot \text{kg}}{\text{m} \cdot \text{Mpc}^2}.$$

Using the conversions: 1 Mpc $= 3.09 \times 10^{22}$m, so that 1Mpc$^2 = 9.55 \times 10^{44}$m^2, and 1 kg $= 10^3$ gm, we find that[13]

$$\rho_c = \frac{87.73 \times 10^{17} \cdot \text{kg}}{\text{m} \cdot \text{Mpc}^2} \times \frac{10^3 \text{gm}}{\text{kg}} \times \frac{1 \text{Mpc}^2}{9.55 \times 10^{44} \text{ m}^2}$$

$$= 9.2 \times 10^{-24} \frac{\text{gm}}{\text{m}^3} = 9.2 \times 10^{-30} \frac{\text{gm}}{\text{cm}^3}.[13]$$

To get some idea how little matter this actually is, a hydrogen atom has a mass of 1.67×10^{-24} gm, so we find that the critical density $\rho_c \approx 6$ hydrogen atoms per cubic meter, which is not a lot, to be sure! Note that his calculation very much depends on the value taken for the Hubble constant H_0 so that values can vary a little, and of course there are rounding off differences.

---

[12] We are only using two decimal places in this calculation due to the uncertainty in the value of the Hubble constant H_0.
[13] This also equates to ~10^{-29}gm/cm^3.

Exercise Work out the critical density taking a value for the Hubble constant $H_0 = 73$ km/s/Mpc. *Ans.* ~9.99×10^{-30} gm/cm^3 ≈ 10^{-29} gm/cm^3.

Exercise Taking the radius of the observable Universe to be 46.5 billion light-years and its mean density to be 10^{-29} gm/cm^3, determine the total mass of the Universe. *Ans.* ~3.6×10^{54} kg.

Density Parameter

But of course, the critical density depends very much on the value of the Hubble Parameter whose current value is still under debate. On the other hand, various attempts have been made to determine the *actual density* of the Universe (see below). A convenient way to discuss this matter is to compare the matter-energy density, $\rho = \rho_{total}$, with the critical density ρ_c. This is expressed in the density parameter Ω defined by the ratio,

$$\boxed{\Omega = \frac{\rho}{\rho_c} = \frac{8\pi G}{3H_0^2}\rho}. \tag{10.16}$$

This means that in the case $k = 0$, by Eq. (10.15) the measured density $\rho = \rho_c$ and therefore $\Omega = 1$ if the geometry is Euclidean.

In the case $k > 0$, then by the Friedmann equation (10.7) we write (taking $c = 1$)

$$\frac{8\pi G}{3}\rho = H_0^2 + \frac{k}{a^2} > H_0^2,$$

so that

$$\rho > \frac{3H_0^2}{8\pi G} = \rho_c$$

which means that $\Omega > 1$.

Likewise, if $k < 0$,

$$\frac{8\pi G}{3}\rho = H_0^2 + \frac{k}{a^2} < H_0^2$$

and it follows that $\rho < \rho_c$ and $\Omega < 1$.

Exercise Conversely, show that if $\Omega = 1$, $\Omega > 1$, $\Omega < 1$ implies that $k = 0$, $k > 0$, $k < 0$, respectively.

From of the preceding exercise, we conclude that, *the characterizations of the geometry of the Universe in terms of the curvature constant when $k = 0$, $k > 0$, $k < 0$, are equivalent to the three cases of the density parameter $\Omega = 1$, $\Omega > 1$, $\Omega < 1$*

respectively. In particular, the latter representations depend upon whether the mass-energy density is equal to the critical density, or is greater than the critical density, or is less than the critical density.

We have seen that the geometry of the Universe is flat at the critical density $\Omega = 1$. This implies an infinite Universe that expands in all directions (Fig. 10.1 left) assuming a likely topology (see sequel) of a 3-dimensional Euclidean space.

If, however, there is more matter/energy in the Universe than given by the critical density, that is $\Omega > 1$ ($k > 0$), then the geometry is that of a sphere,[14] where large triangles in space have greater than 180 degrees. In this case the Universe is known as *closed and finite*. Light rays shining in opposite directions will eventually return to the same point and initially parallel light beams will eventually converge. See Fig. 10.1 center.

Finally, if there is insufficient matter/energy in the Universe so that $\Omega < 1$ ($k < 0$), then the geometry is *open* (hyperbolic)[15] and triangles in space will have angles summing to less than 180 degrees. Initially parallel light beams sent into space will diverge as in Fig. 10.1 right.

Space is Flat

As mentioned above, there have been attempts to measure the matter-energy density of the Universe. One such attempt was the Wilkinson Microwave Anisotropy Probe (WMAP) launched by NASA in 2001 and operated until 2010. It analyzed the Cosmic Background Radiation (CMB) and measured small temperature variations in this faint afterglow of the Big Bang. Another was the Planck Space Telescope operated by the European Space Agency from 2009 until 2013. It also mapped the CMB but had higher precision and sensitivity. A consequence of both studies found that the Universe consisted of 4.9% ordinary matter, 26.8% dark matter, 68.3% dark energy and that $\Omega = 1$ up to a small margin of error of 0.4%. From this we can conclude that the geometry of the Universe is apparently Euclidean, or *flat*, in the terminology of Cosmologists.

Therefore, to express the above in a succinct mathematical form we write

$$\Omega = \Omega_m + \Omega_\Lambda \tag{10.17}$$

where Ω_m represents the contribution of baryonic matter and dark matter, and Ω_Λ the dark energy contribution with (rounded-off) values of $\Omega_m = 0.3$ and $\Omega_\Lambda = 0.7$ (or better, $\Omega_m = 0.317$ and $\Omega_\Lambda = 0.683$).[16]

---

[14] This is actually the surface of a 3-dimensional sphere (*hypersphere*) whereas a normal sphere in space is a 2-dimensional surface. But the analogy holds.

[15] Again, a 3-dimensional hyperboloid but analogous to the usual Euclidean hyperboloid.

[16] Unless discussing the early Universe, the radiation density is negligible and can be excluded.

This is the part of framework of the ΛCDM (lambda cold dark matter) current model of Cosmology.

Let us compare the two densities ρ_Λ (from Eq. (10.6) and ρ_c which gives (again with $c = 1$)

$$\frac{\rho_\Lambda}{\rho_c} = \frac{\Lambda}{8\pi G} \bigg/ \frac{3H_0^2}{8\pi G} = \frac{\Lambda}{3H_0^2}.$$

Moreover, from Eq. (10.16), $\Omega = \frac{8\pi G}{3H_0^2}\rho$, and substituting for our value of ρ_Λ, we can write

$$\Omega_\Lambda = \frac{8\pi G}{3H_0^2}\rho_\Lambda = \frac{\Lambda}{3H_0^2},$$

which from the preceding equation shows that

$$\Omega_\Lambda = \frac{\rho_\Lambda}{\rho_c} \approx 0.7.$$

Consequently, the dark energy density component of the Universe is given by an equivalent mass of: $\rho_\Lambda \approx 0.7 \times 10^{-29} \frac{gm}{cm^3}$, and represents approximately 70% of the mass-energy component of the Universe.

Exercise What is the mass of ordinary matter in the Universe? *Ans.* ~1.76×10^{53} kg.

As an aside it should be mentioned how the Big Bang can occur in an infinite Universe, for if the Universe is infinite that means it has always been so. The main misconception about the Big Bang is that it happened at a specific point in 'space' and spreads into some kind of void. But the proper way to look at it is that the Big Bang happened everywhere all at once and that resulted from a very dense hot state and has expanded into a cooler less dense state that we experience now.

Yet, how can something already infinite expand? As a heuristic illustration, take Hilbert's Infinite Hotel that has rooms numbered 1,2,3,... [17] and suppose that they are all full. A new visitor arrives and wants a room. So, what the hotel manager does is simply move the person in room 1 to room 2 and the person who was in room 2 into room 3 and so on. Then he puts the new visitor in room 1! Everyone now has a room and the manager can actually accommodate a (countably) infinite number of new guests in this manner. This is the difference between the infinite and the finite.

---

[17] This type of infinity where each item can be labelled with a whole number is known as a *countable infinity*. The number of points on a line of any length is a different kind of infinity known as a *continuum*. They are not the same and the continuum is 'greater' than a countable infinity in an appropriate mathematical sense. In fact, there is a hierarchy of increasing infinities and one needs to be very careful when talking about an infinite number of anything.

However, there is a small problem with the experimental data. The WMAP data gives a value for the Hubble Constant as $H_0 = 69.3 \pm 0.8$ (km/s/Mpc), assuming the Universe is flat. As well, the Planck Telescope data give a value of $H_0 = 67.4 \pm 0.5$ (km/s/Mpc). On the other hand, various studies using the best astronomical data from Cepheids, Type 1a supernovae, and the TRGB method as discussed in Chap. 6 indicate values in a range of 72 – 76 km/s/Mpc. For example, the result of the SH0ES collaborative study indicated a value of $H_0 = 73.04 \pm 1.04$ km/s/Mpc.[18] This discrepancy is a major concern in Astronomy and is as yet unresolved. Interestingly, a result in the middle of the discrepancy using TRGB data is that of Freedman et al. at $H_0 = 69.6 \pm 1.7$ km/s/Mpc.[19]

On the other hand, such very similar results from such radically different approaches are rather remarkable! But it also means that there will be some discrepancy in any calculation that involves H_0 as various authors take somewhat different values.

Spacetime Metrics

Recall that in Euclidean 3-space, if a smooth arc α is defined by $x = x(t)$, $y = y(t)$, $z = z(t)$, for $a \leq t \leq b$, the length of the arc is given by

$$s = \int_a^b \left(\left(\frac{dx}{dt}\right)^2 + \left(\frac{dy}{dt}\right)^2 + \left(\frac{dz}{dt}\right)^2 \right)^{1/2} dt.$$

This can also be written in a more abbreviated form for the *arc length element* expressed as[20]

---

[18] A.G. Riess et al., A Comprehensive Measurement of the Local Value of the Hubble Constant with 1 km/s/Mpc Uncertainty from the Hubble Space Telescope and the SH0ES Team, *Astrophys. J. Lett.* 934: L7 (2022), 52 pp.

[19] W. Freedman et al., Calibration of the Tip of the Red Giant Branch, *AJ,* 891:57 (2020), 14 pp.

[20] To be clear, ds^2 means $(ds)^2$. The abbreviated form is a consequence of writing for $a < \tau < b$

$$s(\tau) = \int_a^\tau \left(\left(\frac{dx}{dt}\right)^2 + \left(\frac{dy}{dt}\right)^2 + \left(\frac{dz}{dt}\right)^2 \right)^{1/2} dt$$

for any arbitrarily small arc length segment. By the Fundamental Theorem of Calculus

$$\frac{ds}{d\tau} = \left(\left(\frac{dx}{d\tau}\right)^2 + \left(\frac{dy}{d\tau}\right)^2 + \left(\frac{dz}{d\tau}\right)^2 \right)^{1/2},$$

and first squaring both sides and then multiplying each by $d\tau^2$ results in: $ds^2 = dx^2 + dy^2 + dz^2$.

$$ds^2 = dx^2 + dy^2 + dz^2.$$

However, since space and time are combined in Relativity, it is necessary to define what is meant by the distance between two events in this setting. In the simplest case, we have the spacetime *Minkowski metric* given by its arc length element[21]

$$\boxed{ds^2 = -c^2 dt^2 + dx^2 + dy^2 + dz^2} \tag{10.18}$$

where the general coordinates are (ct, x, y, z). The negative sign means that the time factor and spatial factors contribute in an opposite manner and serve to distinguish between the two. The virtue of taking ct instead of just t is that the units of ct are: $\frac{km}{s} \times s = km$, a distance which is compatible with the units in 3D-space.

Equivalently, in spherical polar coordinates (ct, r, θ, ϕ) [22]

$$\boxed{ds^2 = -c^2 dt^2 + dr^2 + r^2 \left(d\theta^2 + \sin^2\theta d\phi^2\right)} \tag{10.19}$$

measuring distances from the origin in the Euclidean sense. This is the framework of Special Relativity in which there is no gravity (spacetime is flat and not curved by matter or energy).

A useful device in dealing with Special Relativity in this scenario is known as a *light cone* diagram (Fig. 10.2). For purposes of visualization, we first must reduce 3D-space into 2D and make the usual z-axis one of time, namely ct. One can imagine at a particular instant of time an 'event' like a light flash at point E spreading out in a circular fashion with time from its source which results in a solid cone of ever-increasing size at time increases. The boundary of the cone is determined by the speed of light, so any event occurring in spacetime must take place inside the cone.

The upper part of the cone ($t > 0$) represents the future of all possible locations and directions that light can travel over time from the event E. It represents all possible locations where signals from E can propagate to. And the lower cone ($t < 0$) represents all paths along which all possible past events could causally influence the state of the current event at E. Consequently, any event outside the light cone cannot affect the event at E nor be affected by it due to the fact that the speed of light cannot be exceeded.

---

[21] This expression can be written equivalently as: $ds^2 = c^2 dt^2 - dx^2 - dy^2 - dz^2$ and just depends on context. Some versions do not include the velocity of light c in the time term taking $c = 1$. Note also that $d(ct) = cdt$ and $d(ct)^2 = c^2 dt^2$.

[22] Here t is time, r is the radial distance from the origin, θ is the polar angle, and ϕ is the azimuth angle.

Spacetime Metrics

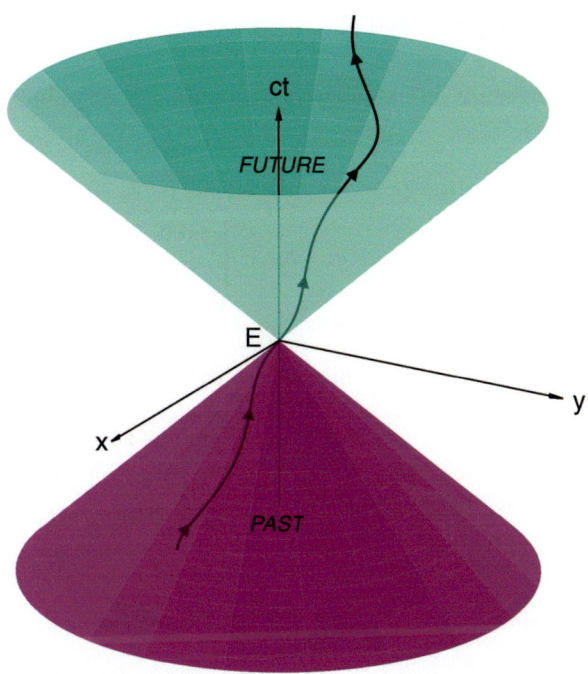

Fig. 10.2 A geometric visualization of the light cone that arises from Special Relativity showing the observer at the origin and the path that an event can take through spacetime. The future light cone extends upwards and the past light cone extends downwards and the position of E is 'now'. The speed of light determines the boundary of the cones in spacetime which are at 45° by convention. (Courtesy Jonathan Park/GPT4)

On the other hand, in General Relativity where matter and energy are taken into account, the resulting spacetime is not flat. Then the infinitesimal distance between two points will be given by the equation

$$\boxed{ds^2 = \sum_{\mu=0}^{4} \sum_{\nu=0}^{4} g_{\mu\nu} dx^\mu dx^\nu} \qquad (10.20)$$

where $g_{\mu\nu}$ is called a *metric tensor* and can be thought of as a matrix whose entries vary according to the local geometry of spacetime. This expression looks rather intimidating but we can dissect it into its components readily enough so let us persevere.

A point in 4-dimensional spacetime has coordinates designated by (x^0, x^1, x^2, x^3) where $x^0 = ct$, and spatial differentials $dx^0 = d(ct)$, dx^1, dx^2, dx^3. The reader must forgive the use of superscripts for coordinate designations which might look confusing but this is standard. The metric tensor is also *symmetric*, which is to say, $g_{\mu\nu} = g_{\nu\mu}$ resulting in only 10 terms that remain.[23]

---

[23] Terms like $g_{01}dx^0 dx^1$ and $g_{10}dx^1 dx^0$ are identical and so are doubled up which leaves only:
$ds^2 = g_{00}(dx^0)^2 + 2g_{01}dx^0 dx^1 + 2g_{02}dx^0 dx^2 + 2g_{03}dx^0 dx^3 + g_{11}(dx^1)^2 + 2g_{12}dx^1 dx^2 + 2g_{13}dx^1 dx^3 + g_{22}(dx^2)^2 + 2g_{23}dx^2 dx^3 + g_{33}(dx^3)^2$.

Minkowski Metric

For example, the Minkowski metric has the very simple metric tensor, namely the matrix

$$\eta_{\mu\nu} = \begin{pmatrix} -1 & 0 & 0 & 0 \\ 0 & 1 & 0 & 0 \\ 0 & 0 & 1 & 0 \\ 0 & 0 & 0 & 1 \end{pmatrix},$$

from which it follows that ds^2 can be written in the form

$$ds^2 = (d(ct), dx, dy, dz) \begin{pmatrix} -1 & 0 & 0 & 0 \\ 0 & 1 & 0 & 0 \\ 0 & 0 & 1 & 0 \\ 0 & 0 & 0 & 1 \end{pmatrix} \begin{pmatrix} d(ct) \\ dx \\ dy \\ dz \end{pmatrix}$$

and performing the matrix multiplication we obtain

$$ds^2 = -c^2 dt^2 + dx^2 + dy^2 + dz^2.$$

The reader is encouraged to verify the calculation here.

Schwarzschild Metric

The very first solution to the Einstein Field Equations was given by Karl Schwarzschild in 1915, that described the gravitational field outside a *spherically symmetric, nonrotating, uncharged (isolated) mass*.[24] The *Schwarzschild metric* where M is the spherical mass is given in terms of the arc length element by the expression

$$\boxed{ds^2 = -\left(1 - \frac{2GM}{rc^2}\right) c^2 dt^2 + \left(1 - \frac{2GM}{rc^2}\right)^{-1} dr^2 + r^2 (d\theta^2 + \sin^2\theta d\phi^2).}$$

(10.21)

---

[24] Schwarzschild made this remarkable discovery while serving on the Russian front during World War I and communicated the result in a letter to Einstein!

Schwarzschild Metric

The corresponding metric tensor again has all zeros off the diagonal with components

$$g_{\mu\nu} = \begin{pmatrix} -\left(1 - \frac{2GM}{rc^2}\right) & 0 & 0 & 0 \\ 0 & \left(1 - \frac{2GM}{rc^2}\right)^{-1} & 0 & 0 \\ 0 & 0 & r^2 & 0 \\ 0 & 0 & 0 & r^2 \sin^2\theta \end{pmatrix},$$

where M is the mass of the object creating the gravitational field and r is the *Schwarzschild radial coordinate*. This latter is somewhat more complex than the simple Euclidean distance from the center due to the curvature of spacetime. Specifically, $r > R_s$ is the circumference divided by 2π of a sphere with the same center as the mass M. The coordinate time t is measured by a very distant observer and not moving with respect to the mass.

The first diagonal term of the matrix, $g_{00} = -\left(1 - \frac{2GM}{rc^2}\right)$, indicates the influence of mass on the curvature of spacetime and how it affects the passage of the *time component* in Eq. (10.21). Moreover, note that the first diagonal term g_{00} actually becomes zero when

$$r = R_s = \frac{2GM}{c^2},$$

which is again the Schwarzschild radius. In this instance the second diagonal term becomes infinite which indicates a coordinate singularity in the metric with the understanding that the metric can no longer describe spacetime at or within this radius. This is the event horizon which is a point of no return for all matter and radiation where the escape velocity equals the speed of light.

In the region where $r < R_s = \frac{2GM}{c^2}$, the Schwarzschild metric is no longer applicable. Where it is applicable, namely outside the Schwarzschild radius[25] with $r > R_s = \frac{2GM}{c^2}$, we then have $\frac{2GM}{rc^2} < 1$, which means that the time-component first term of the Schwarzschild metric

$$-\left(1 - \frac{2GM}{rc^2}\right)c^2 dt^2$$

is always negative. Indeed, $\left(1 - \frac{2GM}{rc^2}\right) = \left(1 - \frac{R_s}{r}\right)$ goes from +1 at an infinite distance to 0 at $r = R_s$. Therefore, as r becomes smaller and approaches the event horizon, the time dilation factor[26] is shrinking to zero. What this means from the perspective of a

---

[25] It is assumed that there are no other masses or forces in the vicinity of the mass M.

[26] Recall from Eq. (8.3) that the gravitational time dilation factor is: $\sqrt{1 - \frac{2GM}{rc^2}} = \sqrt{1 - \frac{R_s}{r}}$.

very distant observer is that the time dilation they observe is becoming infinite compared to the time experienced by an observer very near the Event Horizon where time is effectively standing still, an extreme example of *gravitational time dilation*.

The beauty of the Schwarzschild metric solution is its simplicity.

Note also that for very large values of r, that is $r \gg R_s$, the value of $\left(1 - \frac{R_s}{r}\right) \to 1$ meaning that the Schwarzschild metric is approaching the form of the Minkowski metric of Eq. (10.19) which represents a gravity-free environment.

De Sitter Metric

Another metric came out of this period (1916–1917), and that is the one given by Dutch physicist Willem de Sitter. This is a flat Universe ($k = 0$) in which there is no mass or radiation and is the general assumption that we adopted when considering the dark energy dominated era that was beginning about 9.8 billion years after the Big Bang. The *de Sitter metric* is given by

$$ds^2 = -\left(1 - \frac{\Lambda r^2}{3}\right)c^2 dt + \left(1 - \frac{\Lambda r^2}{3}\right)^{-1} dr^2 + r^2\left(d\theta^2 + \sin^2\theta d\phi^2\right) \quad (10.22)$$

where $\Lambda > 0$ is the cosmological constant.

In this case the density of matter in the Universe is $\rho = 0$ and from the first Friedmann equation this resulted in the scale factor having exponential growth

$$a(t) = e^{H_0 t}.$$

This expansion is purely driven by the positive cosmological constant Λ. The de Sitter Universe is important as dark energy is believed to be driving the current expansion of the Universe.

A few years after the discovery of the Schwarzschild metric came the metric solution for a spherically symmetric, *charged*, non-rotating mass by Reissner and Nordström. While theoretically possible, bodies such as neutron stars or black holes in the Universe are thought to have negligible electric charge.

On the other hand, the *Kerr metric* (1963) describing spacetime outside a *rotating*, uncharged, axially symmetric (that is, symmetric about a particular axis like the axis of rotation) mass is rather more complex and will not be dealt with here.[27] However, it is more realistic than the Schwarzschild metric as virtually all

---

[27] We remark that the Kerr metric includes a term $a = J/Mc$, where J is the angular momentum, and M is the mass, c the velocity of light. Taking $a = 0$ if there is no rotation reduces the Kerr metric to the Schwarzschild metric. The Event Horizon of a Kerr rotating black hole is also dependent on the value of a, namely its radius is given by $R_+ = \frac{GM}{c^2} + \sqrt{\left(\frac{GM}{c^2}\right)^2 - a^2}$ which also reduces to the Schwarzschild radius when $a = 0$.

black holes are thought to be rotating to some extent, and it also describes a framework for understanding other phenomena such as *frame dragging* by which the spacetime geometry is twisted around in the direction of the rotation of a mass such as a black hole or even the Earth for that matter. Regarding the Earth, the frame dragging effect was confirmed by the LAGEOS satellites and Gravity Probe B earlier this century. There is also the *Kerr-Neuman metric* that incorporates electric charge of the rotating mass.

Friedmann-Lemaître-Robertson-Walker Metric

The preceding models all take into account the extreme curvature of spacetime due to a black hole. But in order to describe the large-scale structure of spacetime, cosmologists use the *Friedmann-Lemaître-Robertson-Walker metric*. This solution to the Einstein Field Equations describes a homogeneous and isotropic Universe and is the basis for the Standard Model of Cosmology. Recall, homogeneous means that each location of the Universe has essentially the same physical properties when viewed on a sufficiently large scale and isotropic means that the Universe appears the same in all directions, again when viewed on sufficiently large scale. Taken together, this is known as the *Cosmological Principle*.

The FLRW metric takes into account the curvature constant k which is taken to be either $0, +1, -1$ corresponding to a flat, closed, or open Universe as discussed above, giving

$$ds^2 = -c^2 dt^2 + a(t)^2 \left(\frac{dr^2}{1-kr^2} + r^2 (d\theta^2 + \sin^2\theta d\phi^2) \right), \tag{10.23}$$

where another new feature here is $a(t)$, our scale factor also from the preceding discussion. Furthermore, the r coordinate in the context represents a *comoving distance* measure from an arbitrary point in space and can be chosen as the origin for a coordinate system due to the isotropic assumption about the Universe.

The notion of comoving distance is an integral part of cosmology and factors out the movement of bodies due to the Hubble flow. A body only moving with respect to the Hubble flow has a constant comoving coordinate r and comoving distances between celestial bodies remains constant over time. This allows for the comparisons of distances between bodies at different times in the history of the Universe.

In order to determine the actual *proper distance* at a given time, we have the relation

$$D = a(t) \cdot r$$

with the scale factor being $a(t_0) = 1$ for the present era $t = t_0$ and the time coordinate is $t = 0$ at the time of the Big Bang. That means that the proper distance and comoving distance are the same at present. The motion of a body with respect to the comoving frame is called the *peculiar velocity* and can be due to the gravitational influence of nearby bodies such as other galaxies. One can think of a comoving coordinate system as an expanding grid whereby the coordinate values remain the same and thus the coordinates of objects within the system remain the same unless they have a peculiar motion.

A good example is the movement of the Andromeda and Milky Way galaxies moving towards one another due to their mutual attraction in spite of the Hubble Flow.

In the case of a flat Universe, Eq. (10.23) becomes even simpler,

$$\boxed{ds^2 = -c^2dt^2 + a(t)^2\left(dr^2 + r^2\left(d\theta^2 + \sin^2\theta d\phi^2\right)\right)}. \tag{10.24}$$

So, from this expression it is evident that the scale factor $a(t)$ is the predominant feature in determining the behavior of spacetime in the large in the context of a flat Universe, which seems to be the one we live in.

While both the Special and General Theory of Relativity are stunningly marvelous in describing the workings of the Universe, we must not be led into thinking that they *are* reality. In the words of distinguished physicist N. David Mermin, "Space and time and spacetime are not properties of the world we live in but concepts we have invented to help us organize classical events." By a classical event he means things like: *The train will arrive at 7 pm*. ... The quote by Einstein at the beginning of Chap. 8 says much the same thing.

Topology of the Universe

A likely shape (that is, the *topology*) of the Universe is just a mundane 3-dimensional Euclidean space that is infinite in extent. Space is flat and the geometry is standard Euclidean. But as we have seen, if the curvature $k < 0$, then we could have an open, infinite, saddle shaped Universe in which the angles of a triangle would be less than 180°. However, it is also possible and certainly more economical to create a finite Universe such as the finite closed sphere in which the angles of a triangle would be greater than 180°.

However, there are other more exotic possibilities that are compatible even with a flat Universe. For example, the geometry on a 2-dimensional *torus* (that is, the surface of a donut) is Euclidean in that parallel lines never meet as in Fig. 10.3. Indeed, in 1984 two Russian scientists Alexei Starobinsky and Yakov Zel'dovich, mathematically described a 3-dimensional torus as a possible shape although this is not possible to visualize. There would be no boundaries of such a Universe and it would be finite in extent. This is not just a purely theoretical consideration but there

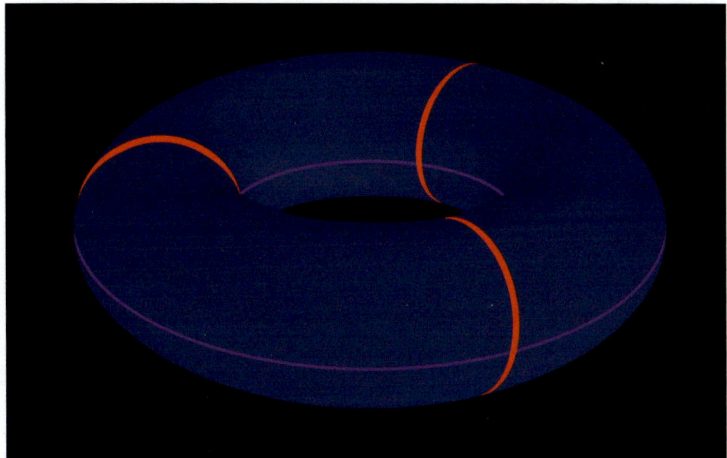

Fig. 10.3 A 2-dimensional torus surface is finite in extent and has the property that parallel lines never meet. Could a 3-dimensional version be the shape of our Universe? (Courtesy Jonathan Park)

is an active group known as *Compact* (Collaboration for Observations, Models and Predictions of Anomalies and Cosmic Topology) that have been sifting through the tea leaves of the CMB (Cosmic Microwave Background radiation) looking for evidence found in minute temperature variations. None of the cosmological evidence thus far has ruled out a more exotic topology for the Universe and at this stage numerous possibilities are being explored.

Other exotic shapes have also been proposed but the true shape of the Universe is unknown. In addition to these topological considerations for our Universe, some scientists have proposed a multitude of universes, the notion of a *multiverse* but all this remains speculation at the present time.

Future of the Universe

We have already discussed how the Universe has gone through various epochs, from radiation dominated, to mass dominated, to the present dark energy dominated era. The ultimate fate of the Universe has various proposed scenarios. It very much depends on factors such as the density, the rate of expansion and the nature of dark energy. None of them look encouraging for our long-term prospects unless there is a change in our present circumstances. But let us look at some of the more popular outcomes nonetheless.

Equation of State Parameter

The fate of the Universe rests upon a simple relationship, namely the *equation of state parameter*

$$\boxed{w = \frac{p}{\rho}} \qquad (10.25)$$

where p is the (negative) pressure of dark energy, and ρ is the energy density (amount of energy per specific volume of space) and we assume $c = 1$ for convenience. The pressure and density can be applied separately to each mass-energy component making up the Universe, such as baryonic matter, dark matter, radiation, and dark energy.

Exercise Use Eq. (10.25) to relate the mass-energy density parameter ρ to the scale factor $a(t)$ via Eq. (10.14) taking $c = 1$ to deduce that: $\rho \propto a^{-3(1+w)}$. Hint: Eliminate p and solve the differential equation for $d\rho/dt$ by separating the variables.

Exercise Via the 2nd Friedmann equation (10.12) with $c = 1$ and dark energy negligible ($\Lambda = 0$) show that the acceleration $\ddot{a} > 0$ implies that $w < -1/3$.

Exercise Using Eq. (10.14) with $c = 1$, show that a constant dark energy density (with time) implies that $p_\Lambda = -\rho_\Lambda$. See the case $w = -1$ below.

Here are various proposed scenarios for how it all ends for a flat Universe ($k = 0$):

$w > -1/3$: In this range the expansion of the Universe will decelerate. In this instance, dark energy does not provide sufficient repulsive pressure to overcome gravitational forces, resulting in a decelerating expansion, yet the Universe would continue to expand indefinitely.

$w = -1/3$: This is a transition value whereby the negative pressure of dark energy counterbalances the gravitational pull which results in a steady rate of expansion that is neither accelerating nor decelerating but increasing at a constant rate.

$-1 < w < -1/3$: In this scenario, the expansion rate of the Universe is accelerating and the ultimate fate of the Universe is a *Heat Death* or *Big Freeze* in which the Universe will gradually cool and galaxies will drift further and further apart as all stars will eventually use up all their fuel and burn out. Even black holes will evaporate over vast time scales and the Universe will be effectively dead. This is the ultimate playing out of the 2nd Law of Thermodynamics with entropy (disorder) increasing to a maximum. It does seem like a great waste to end this way, however.

$w = -1$: This the scenario whereby the cosmological constant reigns supreme and the expansion of the Universe is continually accelerating in an exponential manner again resulting in the Heat Death fate. In this instance, $p = p_\Lambda = -\rho_\Lambda$.

$w < -1$: This is the so-called *phantom energy* regime with the Universe continuing to accelerate under the influence of an extreme form of dark energy and the dark energy density is itself increasing over time, enough so that the dark energy force would overcome all other forces and the fabric of spacetime would be torn apart as well as all galaxies, stars, and even atoms. This is known as the *Big Rip* and would occur after a finite time.

By way of contrast with the preceding scenarios, in the event of a closed universe (k > 0) or open universe (k < 0) which no longer seem so likely, the same values of w would still be relevant to the dynamics of the Universe but the additional feature of curvature would modify the expansion aspects and possibly the ultimate fate. For example, in a closed universe if dark energy is not sufficient the Universe could recollapse in a Big Crunch, whereas if the Universe is open the negative curvature would lead to continued expansion.

A *Big Bounce* scenario is a theoretical consideration whereby the Universe undergoes cyclic expansion and contraction, with each contraction followed by a bounce back into an expansion phase.

The evidence thus far points to a value of $w \approx -1$ and $k \approx 0$ ($\Omega \approx 1$) which leads to the Heat Death scenario as the most likely fate of the Universe in the very long term.

It should be noted that in quintessence and other dark energy models, a modification of the equation of state scenarios is required. In quintessence models for example, the equation of state parameter w evolves with time as the Universe expands, and this would have significant implications for the future of the Universe. This part of the cosmological story has some way to play out yet (Fig. 10.4).

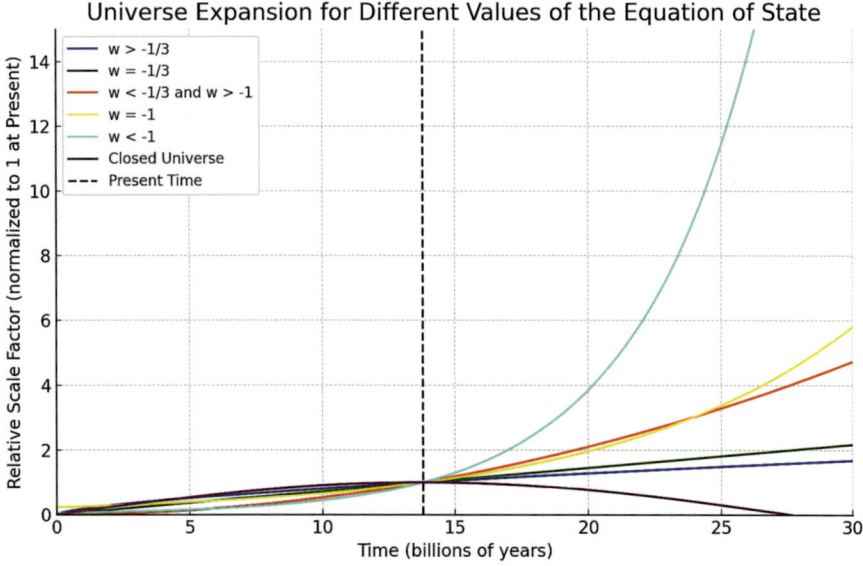

Fig. 10.4 Graph approximating possible scenarios for the ultimate fate of our Universe since the Big Bang. The purple curve at the bottom represents a closed Universe and the Big Crunch outcome. At the other extreme we have the accelerating expansion of the Universe in pale blue, and a decelerating Universe in dark blue. In between, we have the green curve exhibiting a constant expansion rate and the red and yellow curves indicating expansion at an ever-increasing rate. However, current indications are that the Universe continues to expand forever towards the Big Freeze finale. (Courtesy the author and GPT4)

Chapter 11
Epilogue: Are We Alone?

Speculation that we are not alone in the Universe goes back to ancient times with the Greek philosopher Epicurus (341 – 270 BCE) arguing that there were infinitely many worlds, some of which would be like our own. Even in more recent times, there was the 1877 discovery of *canali* (channels) on Mars by the Italian astronomer Giovanni Schiaparelli, a word that was mis-interpreted as canals implying artificial structures. This led to considerable speculation that there was intelligent life on the red planet, an idea especially popularized by the American astronomer Percival Lowell in several books even into the twentieth century.[1]

Drake Equation

The next interesting development along these lines came with the formulation by American astronomer Frank Drake of the famous Drake Equation that was an attempt to estimate how many civilizations exist in the Milky Way that were advanced enough to communicate with us. The equation itself can be broken down into its various components of which there are many:

$$\boxed{N = R_* \times f_p \times n_e \times f_l \times f_i \times f_c \times L}, \tag{11.1}$$

where,

$N = $ The number of civilizations in our galaxy that are able to emit communication signals;
$R_* = $ The average rate of star formation the Milky Way;
$f_p = $ The fraction of those stars that have planetary systems;

---

[1] For example, *Mars and Its Canals*, Macmillan, 1906, which is still available on Amazon.

n_e = The average number of those planets that can potentially support life per star that has planets;

f_l = The fraction of those planets that support life that actually develop life at some stage;

f_i = The fraction of planets with life that go on to develop intelligent life (civilizations);

f_c = The fraction of civilizations that go on to develop technology suitable to send signals into space;

L = The length of time for such civilizations to transmit detectable signals into space.

That is quite some formula! The main difficulty with it is that the various terms have to be estimated based on guesses as most of the parameters are not at all well constrained. The recent detection of thousands of exoplanets has helped pin down estimating the parameters of f_p and n_e. At this stage the reader is wondering what is a reasonable value for N and so is the author so let's make some estimates which are of course subject to great uncertainty:

R_* = The average rate of star formation the Milky Way is estimated to be equivalent to $4 - 8$ solar masses per year.[2] Most new stars are smaller than the Sun at about $0.5\ M_\odot$ but let us take a value of four new stars to be conservative;

f_p = The fraction of those stars that have planetary systems is close to 1, say 0.9 based on exoplanet data;

n_e = The average number of those planets that can potentially support life per star that has planets is very uncertain but estimates vary between 1 and 5, so let us take a value of 3;

f_l = The fraction of those planets that support life that actually develop life at some stage is again highly uncertain with estimates between 10^{-5} and 0.3 so let's take 0.01;

f_i = The fraction of planets with life that go on to develop intelligent life (civilizations) is highly speculative so a conservative estimate would be 0.1;

f_c = The fraction of civilizations that go on to develop technology suitable to send signals into space is another guess as we only have our own civilization to go by so let's take 0.1 again.

L = The length of time that such civilizations transmit detectable signals into space could be from 100 to 10,000 years so let's take 1000.

Therefore, taking the preceding values we obtain

---

[2] T. Siegert et al., Galactic Population Synthesis of Radioactive Nucleosynthesis Ejecta, *A&A* 672, A54 (2023), 1–19.

Fig. 11.1 The magnificent spiral galaxy M83. Does it have inhabitants looking at an image of the Milky Way and wondering if there is any life there? (Courtesy NASA, ESA and the Hubble Heritage Team (STScI/AURA); Acknowledgment: W. Blair (STScI/Johns Hopkins University) and R. O'Connell (University of Virginia))

$$N = 4 \times 0.9 \times 3 \times 0.01 \times 0.1 \times 0.1 \times 1000 \approx 1$$

civilization in our Milky Way that is transmitting signals. That we have not heard any thus far is a bit of a worry,[3] but of course many of the values are highly uncertain so we cannot expect the Drake Equation to give us any definitive answers. However, it does provide us with a framework in which to think about the problem. Furthermore, if a signal has been recently sent from the other side of the Milky Way, 50,000 ly away, then we just have to be patient and await its arrival. And due to the uncertainty in the values chosen above, there could be thousands of advanced civilizations as one might expect in large spiral galaxies such as our own, or M83 (Fig. 11.1), or none whatsoever.

There is also the issue of the existence of advanced civilizations suitably overlapping in time with our own so that a signal sent thousands of years ago will reach us in due course. While some civilizations might still be back in the Stone Age,

---

[3] The *Fermi Paradox* is essentially the question: "If life is plentiful then where are they?"

Fig. 11.2 'Is anyone out there?' (Created by the author with Midjourney)

others could have come and gone already but sent us a signal that is still on its way to us. Will our civilization even last another 50,000 years? Of course, with the estimated two trillion galaxies in the Universe, that does increase the odds significantly for the entire Universe. But the vast distances again become a serious issue for making any form of contact so that we could effectively be alone (Fig. 11.2).

Appendixes

Appendix I

Method of Least Squares/Linear Regression

We suppose that we have data points $(x_1, y_1), (x_2, y_2), \ldots, (x_n, y_n)$, and wish to find the 'best-fit' straight line given by $y = ax + b$ for this data. To this end, as mentioned in Chap. 1, we consider the square of the differences between the y-coordinate of the data points and the y-coordinate of the straight line which is given by $d_i^2 = (ax_i + b - y_i)^2$. This eliminates whether or not the point is above or below the line. Then we minimize the sum

$$E = \sum_{i=1}^{n} d_i^2 = \sum_{i=1}^{n} (ax_i + b - y_i)^2$$

which is a function of the two *variables a, b*.

In order to solve for a and b we have to differentiate the quantity E with respect to a and set that equal to zero and then do the same for b. In both cases, the values of x_i and y_i are constants. It is much easier than it looks since keeping b fixed

$$\frac{\partial E}{\partial a} = \sum_{i=1}^{n} 2x_i(ax_i + b - y_i) = 0.$$

This means that

$$\sum_{i=1}^{n} 2ax_i^2 = \sum_{i=1}^{n} 2x_i y_i - \sum_{i=1}^{n} 2x_i b,$$

and cancelling the 2s and solving for a,

$$a = \frac{\sum_{i=1}^{n} x_i y_i - \sum_{i=1}^{n} x_i b}{\sum_{i=1}^{n} x_i^2}. \tag{A1}$$

This still has a b term in it but we will now differentiate with respect to b keeping a fixed and we will end up with two equations in two unknowns. Thus

$$\frac{\partial E}{\partial b} = \sum_{i=1}^{n} 2(ax_i + b - y_i) = 0,$$

implying that

$$nb = \sum_{i=1}^{n} y_i - a \sum_{i=1}^{n} x_i$$

and therefore,

$$b = \frac{\sum_{i=1}^{n} y_i - a \sum_{i=1}^{n} x_i}{n},$$

which is our second equation involving a and b.

Now we simply have to go back and put this value of b into Eq. (A1) to give

$$a = \frac{\sum_{i=1}^{n} x_i y_i - \sum_{i=1}^{n} x_i b}{\sum_{i=1}^{n} x_i^2}$$

$$= \frac{\sum_{i=1}^{n} x_i y_i - \sum_{i=1}^{n} x_i \left(\frac{\sum_{i=1}^{n} y_i - a \sum_{i=1}^{n} x_i}{n} \right)}{\sum_{i=1}^{n} x_i^2}$$

$$= \frac{n \sum_{i=1}^{n} x_i y_i - \left(\sum_{i=1}^{n} x_i \right) \left(\sum_{i=1}^{n} y_i \right) + a \left(\sum_{i=1}^{n} x_i \right)^2}{n \sum_{i=1}^{n} x_i^2}.$$

Solving this expression for a,

$$a = \frac{n \sum_{i=1}^{n} x_i y_i - \left(\sum_{i=1}^{n} x_i \right) \left(\sum_{i=1}^{n} y_i \right)}{n \sum_{i=1}^{n} x_i^2 - \left(\sum_{i=1}^{n} x_i \right)^2},$$

and b is already given above once we have determined the value of a.

Appendix II

Exponential Growth/Decay

Many quantities in Nature either grow or decay at a rate proportional to the amount of the quantity present at any given time. The more there is the more rapidly it increases or decays. Certain populations of living organisms or radioactivity are good examples of this type of behavior of growth or decay respectively. The latter is the most pertinent for us.

To state this mathematically, let $N(t) \geq 0$ be the amount of a particular quantity present at time t, say a radioactive substance, measured in any units. Since its rate of change which we denote as dN/dt is proportional to the amount $N(t)$, at each time t, we have

$$\frac{dN}{dt} = kN(t).$$

We separate the variables and integrate each side

$$\int \frac{dN}{N} = \int k\,dt,$$

and therefore,

$$\log N(t) = kt + C,$$

so that

$$N(t) = C_1 e^{kt}.$$

At time $t = 0$, then $N(0) = C_1$, which is the initial amount present and we have

$$N(t) = N(0) e^{kt}.$$

When $k > 0$, the equation describes *exponential growth*, and when $k < 0$ we have *exponential decay*.

An important feature of radioactivity is the time for one-half of the substance to decay, called the *half-life*, denoted by $t_{1/2}$. It is readily found by setting $N(t) = \frac{1}{2}N(0)$, giving

$$N(t_{1/2}) = \frac{1}{2}N(0) = N(0)e^{kt_{1/2}},$$

and solving for $t = t_{1/2}$,

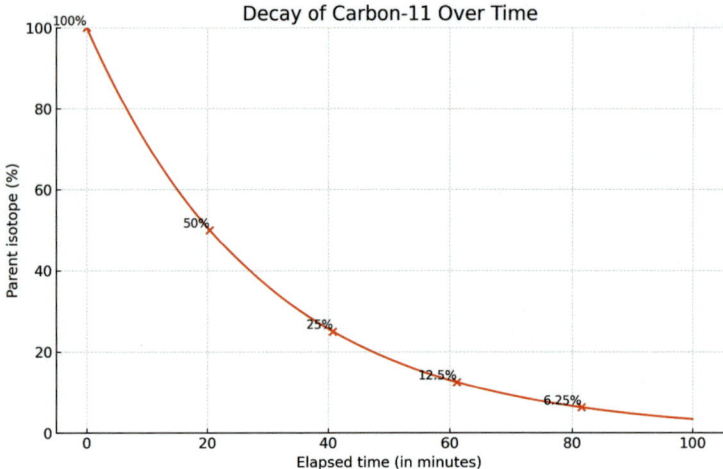

Fig. A1 The radioactive decay of Carbon-11 with a half-life of 20.38 minutes that is used in medical imaging

$$t_{1/2} = -\frac{\ln 2}{k}. \tag{A2}$$

For radioactive substances, this can vary between a few microseconds (Polonium-214; $t_{1/2} = 164$ microseconds) to minutes (Carbon-11; $t_{1/2} = 20.38$ minutes), to billions of years (Uranium-238; $t_{1/2} = 4.468$ billion years, and Thorium-232; $t_{1/2} = 14.05$ billion years). For radioactive substances, once this half-life is known it allows for the determination of the constant k. A consequence of Eq. (A2) is

$$k = -\frac{\ln 2}{t_{1/2}}. \tag{A3}$$

Note that the constant k is negative, indicating decay.

In Fig. A1 we have a graph of Carbon-11 as it decays over time. It is used in positron emission tomography (PET) for medical imaging.

Appendix III

Wien's Displacement Law from Planck's Law

Starting with Planck's Law

$$B(\lambda, T) = \frac{2hc^2}{\lambda^5} \cdot \frac{1}{e^{\frac{hc}{\lambda k_B T}} - 1}$$

let us first simplify this expression by setting $\alpha = 2hc^2$ and since T will be constant, $\beta = \frac{hc}{k_B T}$, so that we now have

$$I(\lambda, T) = \frac{\alpha}{\lambda^5} \cdot \frac{1}{e^{\beta/\lambda} - 1}.$$

In order to find the maximum, we differentiate with respect to λ via the product rule,

$$\begin{aligned}\frac{dI}{d\lambda} &= \frac{\alpha}{\lambda^5} \cdot \frac{-1}{(e^{\beta/\lambda} - 1)^2} \cdot e^{\beta/\lambda} \left(\frac{-\beta}{\lambda^2}\right) + \left(\frac{1}{e^{\beta/\lambda} - 1}\right) \cdot \frac{-5\alpha}{\lambda^6} \\ &= \frac{\alpha}{\lambda^6 (e^{\beta/\lambda} - 1)} \left(\frac{\beta e^{\beta/\lambda}}{\lambda(e^{\beta/\lambda} - 1)} - 5\right) = \frac{\alpha}{\lambda^6 (e^{\beta/\lambda} - 1)^2} \left(\frac{\beta e^{\beta/\lambda}}{\lambda} - 5 e^{\frac{\beta}{\lambda}} + 5\right).\end{aligned}$$

We next set the derivative to zero and since the first term is nonzero,

$$\frac{dI}{d\lambda} = \frac{\beta e^{\beta/\lambda}}{\lambda} - 5 e^{\beta/\lambda} + 5 = 0,$$

that is,

$$5 e^{\beta/\lambda} - \frac{\beta e^{\beta/\lambda}}{\lambda} = 5.$$

To solve this let us substitute, $x = \beta/\lambda$ so that

$$5 e^x - x e^x = 5$$

and finally,

$$e^x (5 - x) = 5. \tag{A4}$$

The little snag here is that this innocuous expression is actually a *transcendental equation* involving the transcendental function e^x. The solution to the equation can be found by the *Newton-Raphson method* and is: $x = 4.965$ to three decimal places. Then with $\beta = hc/k_B T$ and plugging in the values of h, c, and x, we have the Wien Law

$$\lambda = \frac{\beta}{x} = \frac{hc}{k_B x} \cdot \frac{1}{T} = \frac{0.0028977 \text{ m} \cdot \text{K}}{T}.$$

Here is a Python program for the Newton-Raphson method for Eq. (A4) since every Science student should know how to use it due to its practicality.

```python
import math
def newton_raphson_method(f, df, x0, max_iterations, tolerance):
    x = x0
    iterations = 0
    error = float('inf')

    while error > tolerance and iterations < max_iterations:
        x_previous = x
        x = x_previous - f(x_previous) / df(x_previous)
        iterations += 1
        error = abs(x - x_previous)

    if iterations >= max_iterations:
        return None  # The method did not converge within the maximum number of iterations.
    else:
        return x  # Return the approximate root
def main():
    # Define the function
    f = lambda x: (5 - x) * math.exp(x) - 5
    # Define the derivative of the function
    df = lambda x: (4 - x) * math.exp(x)
    # Set the initial guess
    x0 = 6
    # Set the maximum number of iterations
    max_iterations = 100
    # Set the tolerance for convergence
    tolerance = 1e-6
    # Call the Newton-Raphson method
    root = newton_raphson_method(f, df, x0, max_iterations, tolerance)
    if root is None:
        print('The method did not converge within the maximum number of iterations.')
    else:
        print('The root is approximately {:.6f}'.format(root))
if __name__ == '__main__':
    main()
```

The root is approximately 4.965114

Appendix IV

Wien's Approximation Law

We start with Planck's Law

$$B(\lambda, T) = \frac{2hc^2}{\lambda^5} \cdot \frac{1}{e^{\frac{hc}{\lambda k_B T}} - 1}$$

and simplify the exponential second factor again. To this end, let $\alpha = \frac{hc}{\lambda k_B T}$ so that the exponential part can be written as

$$\frac{1}{e^\alpha - 1} = \frac{1}{e^\alpha} \frac{1}{(1 - e^{-\alpha})}. \tag{A5}$$

From the Taylor series for any positive value $x < 1$, we have

$$\frac{1}{1-x} = 1 - x + x^2 - \ldots$$

Then the Taylor series for x^{-1} whenever $x > 1$ gives

$$\frac{1}{1 - x^{-1}} = 1 + x^{-1} - x^{-2} + \ldots \approx 1 + x^{-1}$$

dropping off the higher order terms.

Setting $x = e^\alpha > 1$ so that $x^{-1} = e^{-\alpha} = e^{-\frac{hc}{\lambda k_B T}}$, we obtain from the preceding approximation

$$\frac{1}{1 - e^{-\alpha}} = \frac{1}{1 - x^{-1}} \approx 1 + x^{-1} = 1 + e^{-\alpha}.$$

We conclude from this expression and Eq. (A5) that

$$\frac{1}{e^\alpha - 1} = \frac{1}{e^\alpha} \frac{1}{(1 - e^{-\alpha})} \approx \frac{1}{e^\alpha}(1 + e^{-\alpha}) = e^{-\alpha} + e^{-2\alpha} \approx e^{-\alpha},$$

in other words,

$$\frac{1}{e^{\frac{hc}{\lambda k_B T}} - 1} \approx e^{-\frac{hc}{\lambda k_B T}}.$$

This gives us the approximation of Wien (Eq. 3.4)

$$B(\lambda, T) \approx \frac{2hc^2}{\lambda^5} e^{-\frac{hc}{\lambda k_B T}}.$$

Appendix V

Stefan-Boltzmann Law

Let us sum up the intensity of a blackbody in terms of frequency instead of wavelength as it is more straightforward, which yields

$$I_0 = \int_0^\infty B(f,T)df,$$

which gives the power per unit solid angle per unit area per unit of time across all frequencies.

In order to compute the total power E in W/m² we need to consider the angular distribution of the radiation as received by a hemisphere above an ideal flat blackbody. Since the hemisphere has 2π steradians one might think that we simply multiply I_0 by 2π. However, the intensity over this hemisphere varies in accordance with Lambert Cosine Law and so we actually have[1]

$$E = \pi I_0 = \pi \int_0^\infty B(f,T)df.$$

Substituting for $B(f,T)$ from footnote 12, Chapter 3,

---

[1] The *Lambert Cosine Law* states that the intensity of radiation emitted in a particular direction is a function of the cosine of the angle between the direction of propagation and the normal to the surface. Mathematically,

$$I(\theta) = I_0 \cos\theta,$$

where I_0 is the maximum intensity and θ is the angle between the direction of the propagation of the radiation and the normal to the emitting surface. So $I(\theta)$ attains its maximum value I_0 at $\theta = 0$ (the straight up direction) and then becomes attenuated as the angle θ increases. Therefore, if we sum up (by integration) the intensity $I(\theta)$ over the entire hemisphere from $\theta = 0$ to $\theta = \frac{\pi}{2}$ and $\phi = 0$ to 2π, we obtain using spherical coordinates and integrating over a *solid angle element* of the sphere $d\Omega = \sin\theta d\theta d\phi$ as in Fig. A2,

$$E = \int_0^{2\pi}\int_0^{\frac{\pi}{2}} I_0 \cos\theta \cdot \sin\theta d\theta d\phi$$

$$= 2\pi\left(\frac{1}{2}\right)I_0 = \pi I_0.$$

$$E = \pi \int_0^\infty \frac{2hf^3}{c^2} \cdot \frac{1}{e^{\frac{hf}{k_B T}} - 1} df = \frac{2h\pi}{c^2} \int_0^\infty f^3 \cdot \frac{1}{e^{\frac{hf}{k_B T}} - 1} df.$$

To simplify matters we make the substitution $x = \frac{hf}{k_B T}$, so that $f = \frac{k_B T}{h} x$, implying that

$$f^3 = \frac{k_B^3 T^3}{h^3} x^3 \text{ and } df = \frac{k_B T}{h} dx.$$

Finally, putting all these ingredients together into the integral and factoring out all the many constants,

$$E = \pi \int_0^\infty B(f,T) df = \frac{2h\pi}{c^2} \cdot \frac{k_B^3 T^3}{h^3} \cdot \frac{k_B T}{h} \int_0^\infty \frac{x^3}{e^x - 1} dx$$
$$= \frac{2\pi k_B^4 T^4}{c^2 h^3} \int_0^\infty \frac{x^3}{e^x - 1} dx.$$

The integral on the right has an exact value[2] which is $\frac{\pi^4}{15}$, so that we arrive at the Stefan-Boltzmann Law Fig. A2

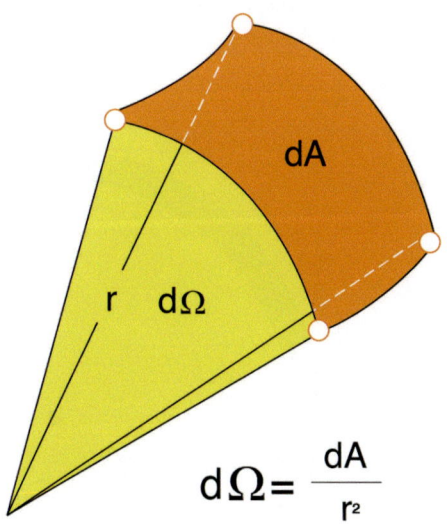

Fig. A2 $d\Omega = \frac{dA}{r^2} = rd\theta \cdot r\sin\theta d\phi / r^2 = \sin\theta d\theta d\phi$. (Courtesy Katy Metcalf)

---

[2] This integral is related to the Riemann zeta function and the Gamma function which are subjects beyond the scope of this book but the calculation of the integral then becomes straight-forward.

$$E = \int_0^\infty B(f,T)df = \frac{2\pi k_B^4 T^4}{c^2 h^3} \cdot \frac{\pi^4}{15} = \left(\frac{2\pi^5 k_B^4}{15 c^2 h^3}\right) T^4 = \sigma T^4.$$

Appendix VI

Gravitational Potential Energy

We will assume that the spherical mass has a uniform density ρ and consider an infinitesimally thin shell around the exterior of the sphere S as in Fig. A3.

We are now in a position to compute the gravitational potential energy of the thin shell which has a mass denoted by dm. In view of Eq. (4.3) we have

$$dU_S = -\frac{GM(r)dm}{r}.$$

Here $M(r)$ is the mass of the sphere inside the radius r and is given by

$$M(r) = \frac{4}{3}\pi r^3 \cdot \rho.$$

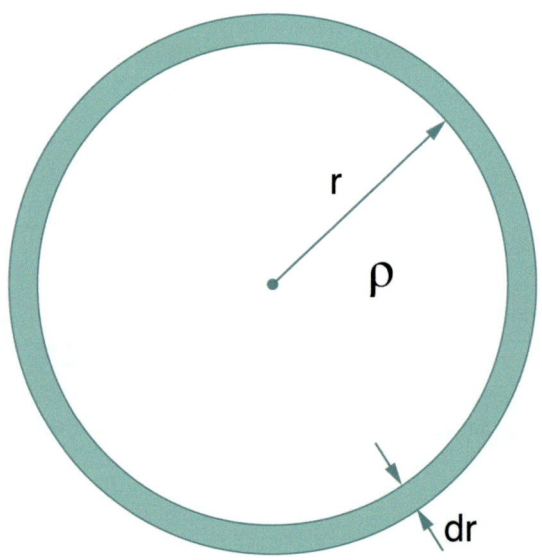

Fig. A3 A sphere of uniform density ρ with a thin shell of thickness dr at a radius r. (Courtesy Katy Metcalf)

As well, the mass *dm* of the thin shell can be expressed as the area times the density,[3]

$$dm = 4\pi r^2 dr \cdot \rho.$$

Putting these last two expressions into the preceding one,

$$dU_S = -\frac{G(\frac{4}{3}\pi r^3 \cdot \rho)4\pi r^2 dr \cdot \rho}{r}$$

so that

$$dU_S = -\frac{16}{3}\pi^2\rho^2 G r^4 dr.$$

To sum up the gravitational potential energy of each shell to obtain the same for the entire sphere *S*, we integrate from 0 to *R*,

$$U_S = -\frac{16}{3}G\pi^2\rho^2 \int_0^R r^4 dr.$$

The integration gives us

$$U_S = -\frac{16}{15}G\pi^2\rho^2 R^5.$$

This result is fine but we can do better and eliminate the ρ^2 term by dividing (and multiplying) this expression by M^2 since for $M = \frac{4}{3}\pi R^3 \cdot \rho$,

$$M^2 = \frac{16}{9}\pi^2 R^6 \rho^2,$$

Therefore,

$$U_S = -\frac{16}{15}G\pi^2\rho^2 R^5 \times \frac{M^2}{\frac{16}{9}\pi^2 R^6 \cdot \rho^2} = -\frac{3}{5}\frac{GM^2}{R},$$

which is the gravitational potential energy of a sphere of uniform density. Note that the density term has now been eliminated so its actual value is of no consequence.

---

[3] To see this one must consider $dm = \rho dV$, where the latter term is given by $dV = \frac{4}{3}\pi(r+dr)^3 - \frac{4}{3}\pi r^3$ and expand the cubic expression and simplify, dropping out the terms involving $(dr)^2$ and $(dr)^3$ which are negligible. Details are left to the reader.

Appendix VII

Python Computer Programs[4]

Gravitational Time Dilation of Earth Clock Per Year

```
import math
# Gravitational constant
G = 6.67430e-11
# Speed of light
c = 299792458
# Mass of Earth (kg)
M = 5.9722e24
# Radius of Earth
R = 6.371e6
# Seconds per year
S = 31557600
# Time dilation factor
T1 = (2 * G * M) / (R * (c ** 2))
T2 = math.sqrt(1 - T1)
# Time dilation factor in seconds
S1 = T2 * S
# Time difference in seconds
S2 = S - S1
print(S2)
0.02197 seconds per year lagging behind a clock in space
```

---

[4] An online python compiler is at: https://www.programiz.com/python-programming/online--compiler/

Gravitational Time Dilation on Sun's Surface Per Year

```python
import math
# Gravitational constant
G = 6.67430e-11
# Speed of light
c = 299792458
# Mass of Sun (kg)
M = 1.9885e30
# Radius of Sun (m)
R = 6.96340e8
# Number of seconds per year
S = 31557600
# Time dilation factor
T1 = (2 * G * M) / (R * (c ** 2))
T2 = math.sqrt(1 - T1)
# Time dilation per year on Sun
T3 = S * T2
# Time difference per year on Sun
T4 = S - T3
print(T4)
66.9226 seconds per year
```

Relativistic Velocity Time Dilation of GPS Satellites Per Day Due to Velocity

import math

Velocity of light

c = 299792458

Velocity of satellite = 14,000 km/hr = 14,000,000 m/hr X 1 hr/3600 s = 3888.88888888 m/s

v = 3888.88888888

Seconds per day

S = 86400

Reciprocal of time dilation factor gives the fractional rate of time lag on satellite

T1 = math.sqrt(1 - (v / c) ** 2)

Dilation per day in seconds

S1 = T1 * S

Number of seconds lagging behind per day

S2 = S - S1

print(S2)

7.2693e-06 seconds per day time lag compared to stationary clock

Relativistic Gravitational Time Dilation of GPS Satellites Per Day Due to Gravity

```
import math
# Gravitational constant
G = 6.67430e-11
# Speed of light
c = 299792458
# Mass of Earth (kg)
M = 5.9722e24
# Radius of Satellite (m) = radius of Earth + altitude of satellite
# = 6371 km + 20,200 km = 26,571 km = 26,571,000 m from the center of Earth
R = 2.6571e7
# Seconds per day
S = 86400
# Time dilation factor
T1 = (2 * G * M) / (R * (c ** 2))
T2 = math.sqrt(1 - T1)
# Dilation per day in seconds
S1 = T2 * S
# Number of seconds lagging behind distant space clock per day
S2 = S - S1
print(S2)
```

1.4421e-05 seconds per day lag behind a distant space clock

Appendix VIII

Apsidal Precession

Einstein originally published his landmark result on the precession of the orbit of Mercury in 1915.[5] The same calculation occurs in Chap. 8 using Eq. (8.4) which is Einstein's equation (75) below in his monumental work on the General Theory of Relativity: *Die Grundlage der allgemeinen Relativitätstheorie* (1916):

Berechnet man das Gravitationsfeld um eine Größenordnung genauer, und ebenso mit entsprechender Genauigkeit die Bahnbewegung eines materiellen Punktes von relativ unendlich kleiner Masse, so erhält man gegenüber den Kepler-Newtonschen Gesetzen der Planetenbewegung eine Abweichung von folgender Art. Die Bahnellipse eines Planeten erfährt in Richtung der Bahnbewegung eine langsame Drehung vom Betrage

$$(75) \qquad \varepsilon = 24\,\pi^3 \frac{a^2}{T^2 c^2 (1-e^2)}$$

pro Umlauf. In dieser Formel bedeutet a die große Halbachse, c die Lichtgeschwindigkeit in üblichem Maße, e die Exzentrizität, T die Umlaufszeit in Sekunden.[1]

Die Rechnung ergibt für den Planeten Merkur eine Drehung der Bahn um 43″ pro Jahrhundert, genau entsprechend der Konstatierung der Astronomen (Leverrier); diese fanden nämlich einen durch Störungen der übrigen Planeten nicht erklärbaren Rest der Perihelbewegung dieses Planeten von der angegebenen Größe.

---

1) Bezüglich der Rechnung verweise ich auf die Originalabhandlungen A. Einstein, Sitzungsber. d. Preuß. Akad. d. Wiss. **47.** p. 831. 1915. — K. Schwarzschild, Sitzungsber. d. Preuß. Akad. d. Wiss. **7.** p. 189. 1916.

(Eingegangen 20. März 1916.)

---

[5] Erklärung der Perihelbewegung des Merkur aus der allgemeinen Relativitätstheorie, *Sitz. König. Preuß. Akad. Wiss.* (Berlin), (1915), 831–839.

Appendix IX

Einstein Ring

Here we analyze the situation of Fig. 8.12 more closely. Using the notation of that figure in the one below, the first trigonometric observation is

$$D_{SI} = (D_{LS} + D_L)\tan\beta,$$

and since the angular radius β will be very small, we can invoke the small angle approximation and hence

$$D_{SI} = (D_{LS} + D_L)\beta. \tag{A6}$$

Likewise,

$$b = D_L \tan\beta = D_L\beta \tag{A7}$$

as in Fig. A4.

Next, we need to consider the angle of deflection α. In dealing with the triangle SaI, we can drop a perpendicular line down to the line SI that has length D_{LS} as in Fig. A5.

Clearly, $\alpha_1 + \alpha_1 = \alpha$ and $L_1 + L_2 = D_{SI}$. Again by the small angle approximation,

$$L_1 = D_{LS}\tan\alpha_1 = D_{LS}\alpha_1,$$

$$L_2 = D_{LS}\tan\alpha_2 = D_{LS}\alpha_2.$$

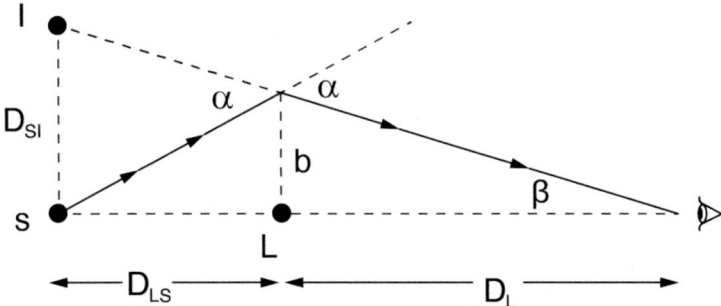

Fig. A4 The angular radius β of the Einstein ring formed from a star source at S in a direct line with the lens mass L and the observer to the right. The path of a light ray reaching the observer is the solid lines with arrows. Only those rays with a certain 'impact parameter' will reach the observer. (Courtesy Katy Metcalf)

Fig. A5 A closer look at the triangle $S\alpha I$ from the preceding figure where $\alpha_1 + \alpha_1 = \alpha$ and $L_1 + L_2 = D_{SI}$. (Courtesy Katy Metcalf)

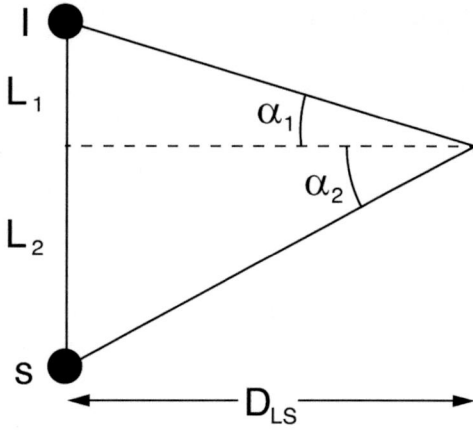

Just adding these two equations yields

$$D_{SI} = D_{LS}\alpha.$$

Invoking Eq. (A6) with this equation gives

$$(D_{LS} + D_L)\beta = D_{LS}\alpha$$

and solving for β,

$$\beta = \frac{D_{LS}\alpha}{D_{LS} + D_L}. \tag{A8}$$

We already know what the angular deflection α is from Eq. (8.10) of Chap. 8: $\alpha = \frac{4GM}{c^2 b}$ and it follow that

$$\beta = \frac{4GM}{c^2 b} \frac{D_{LS}}{D_{LS} + D_L}.$$

Finally, we can substitute the value for b, namely $b = D_L\beta$, implying the angular radius of the ring will be

$$\boxed{\beta = \sqrt{\frac{4GM}{c^2} \frac{D_{LS}}{D_L(D_{LS} + D_L)}}.} \tag{A9}$$

For the physical radius of the ring which is equal to $b = D_L\beta$,

Appendixes

$$b = D_L\sqrt{\frac{4GM}{c^2}\frac{D_{LS}}{D_L(D_{LS}+D_L)}}$$

and we obtain

$$\boxed{b = \sqrt{\frac{4GM}{c^2}\frac{D_L D_{LS}}{(D_{LS}+D_L)}}}. \tag{A10}$$

Appendix X

Multiply-Lensed Objects

Below are a couple of images (Figs. A6 and A7) illustrating the dramatic effects of gravitational lensing.

Fig. A6 Galaxy cluster Abell 370, located about 4 billion light-years away, contains a wide assortment of several hundred galaxies held together by their mutual gravitational attraction. Among the galaxies are arcs of blue light that are distorted images of remote galaxies behind the cluster that are too faint to see directly. Instead, the gravity from the cluster acts as a gravitational lens that magnifies and stretches the images of background galaxies. (Credit: NASA, ESA, and J. Lotz and the HFF Team (STScI))

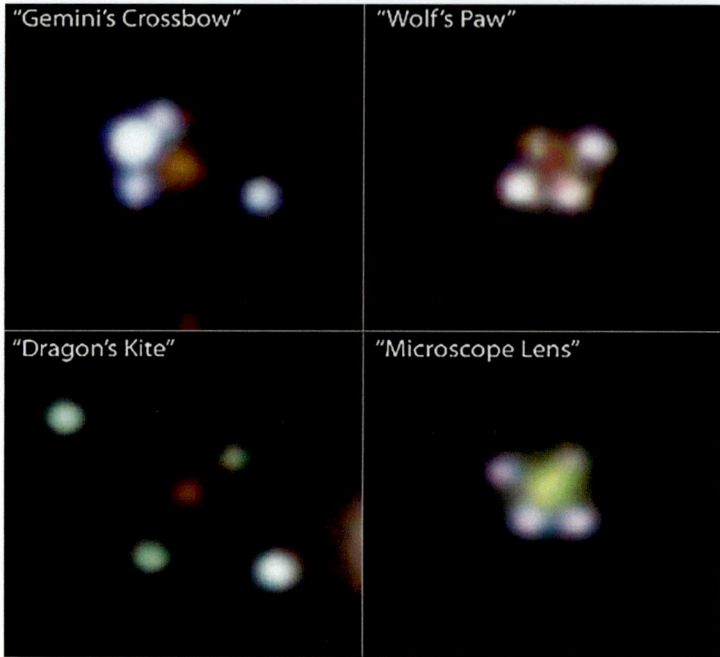

Fig. A7 Four of the newfound quadruply-imaged quasars. Top right and moving counterclockwise, the objects are: GraL J1537-3010 or Wolf's Paw; GraL J0659+1629 or Gemini's Crossbow; GraL J1651-0417 or Dragon's Kite; GraL J2038-4008 or Microscope Lens. The fuzzy dot in the middle of the images is the lensing galaxy whose gravity is splitting the light from the quasar behind it in such a way as to produce four quasar images. The images of Wolf's Paw, Gemini's Crossbow, and Dragon's Kite were taken by the Pan-STARRS1 Sky Survey; and the image Microscope Lens was captured by the Dark Energy Survey. (Image credit: Gaia Gravitational Lenses (GraL) group)

Appendix XI

Gaussian Curvature

The notion of curvature appears in the Friedmann Equations and each value gives a different geometry of the Universe. Let us recall:

$k = 0$: Euclidean (flat) universe;
$k = 1$: Closed universe with positive curvature;
$k = -1$: Open universe with negative curvature.

The constant however, does not give the magnitude of the curvature, just the nature of the curvature itself.

In a mathematical setting, the notion of Gaussian curvature is used which does give the magnitude of the curvature. It is a measure of how much the curve deviates

Fig. A8 The circle that best fits the curve C at the point P and is tangent at P. Public domain

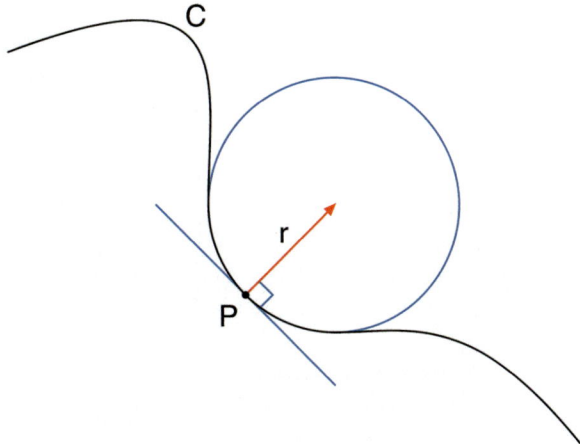

from a straight line. Informally, to define this curvature let us consider any point P on a curve C in the plane. Then define the *curvature* at the point P to be

$$\kappa = 1/r$$

where r is the radius of the *osculating* circle (Fig. A8) that is tangent to the curve at P and best fits the curve at P.

At a point P on a surface in 3-space, there are infinitely many curves on the surface passing through P. One takes the respective maximum and minimum curvature values κ_1, κ_2 (known as the *principal curvatures*) and we take their product to give the *Gaussian curvature* at the point P,

$$\mathcal{K} = \kappa_1 \cdot \kappa_2.$$

The principal curvatures can be positive or negative and always lie in perpendicular directions to one another.[6] If a curve bends away from an outer normal,[7] then the curvature is positive, and if the curve bends in the direction of the outer normal, then the curvature is deemed negative.

For example, on a flat plane, the best fit circles would all have an infinite radius so that $\kappa_1 = \kappa_2 = 0$ and so its Gaussian curvature is zero as one would expect. Considering a sphere of radius R, then all osculating circles have curvature $\kappa = \frac{1}{R}$ implying that $\mathcal{K} = \frac{1}{R^2} > 0$. In the case of a cylinder of radius r, then the maximum curvature would be $\kappa_1 = \frac{1}{r}$ and the minimum curvature is $\kappa_2 = 0$, so that $\mathcal{K} = \kappa_1 \cdot \kappa_2 = 0$. In the case of a hyperboloid, which is a hyperbola rotated around an

---

[6] This is a result of differential geometry and will not be dealt with here.

[7] Generally speaking, an outer normal is perpendicular to the tangent of a curve or surface and points away from the 'inside' of the curve or surface.

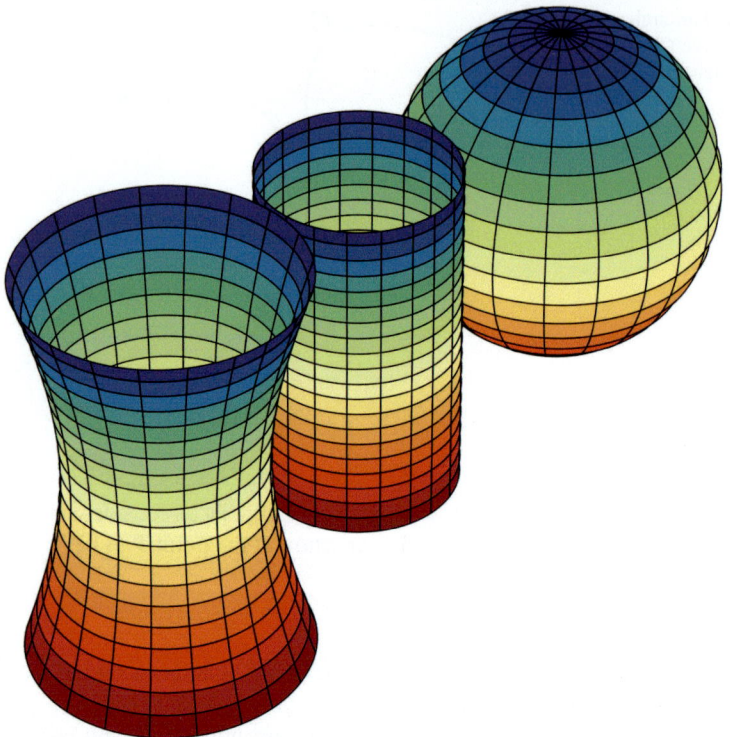

Fig. A9 A hyperboloid with negative Gaussian curvature, a cylinder with Gaussian curvature zero, and a sphere with positive Gaussian curvature. (Public Domain/Nicoguaro)

axis in 3-space, the radius r from axis of rotation forms a circle of radius r so that we have at any point P on the surface, $\kappa_1 = \frac{1}{r} > 0$, since the circle curves away from an outward normal to the surface at P. However, the other principal curvature κ_2 is negative as it curves in the same direction as the outer normal. Then $\mathcal{K} = \kappa_1 \cdot \kappa_2 < 0$ for the hyperboloid and similarly for a saddle shaped surface. Furthermore, a point on the outer surface of a torus has positive Gaussian curvature and negative Gaussian curvature on its inner surface as in Fig. 10.3.

The relation between Gaussian curvature and the Friedmann curvature constant k is given by

$$\mathcal{K} = \frac{k}{a(t)^2}.$$

In this formulation, the Friedmann curvature determines the type of curvature of the Universe and the scale factor determines the magnitude of the curvature as it varies over time.

Therefore:

$k = 0$: implies that the Gaussian curvature $\mathcal{K}$ is also zero, like the plane or cylinder.

$k = 1$, then the Gaussian curvature is positive, and varies inversely with the square of the scale factor. As the scale factor $a(t)$ increases the Gaussian curvature $\mathcal{K}$ decreases.

$k = -1$: then the Gaussian curvature is also negative, like a hyperboloid or saddle.

Example Considering a torus that is formed by the rotation of a circle of radius r whose center is at distance $R > r$ from the axis of rotation. Then for a point P on the outermost equator of the torus, $\kappa_1 = \frac{1}{r}$ and $\kappa_2 = \frac{1}{R+r}$, so that $\mathcal{K} = \frac{1}{r(R+r)}$. And for a point on the innermost equator of the torus, $\kappa_1 = \frac{1}{r}$ and $\kappa_2 = \frac{-1}{R-r}$, giving $\mathcal{K} = \frac{-1}{r(R-r)}$. Moreover, since the Gaussian curvature varies continuously over a smooth surface, and since some points have positive curvature and some negative, then there are points on the torus that have zero curvature!

Appendix XII

Second Friedmann Equation from First Friedmann Equation

We start with the first Friedmann Equation and will set the curvature constant $k = 0$ which slightly simplifies the calculation so that by Eq. (10.5)

$$H^2 = \frac{8\pi G}{3}\rho + \frac{\Lambda c^2}{3}.$$

Differentiating this expression with respect to the variable t via the product rule

$$2H\dot{H} = \frac{8\pi G}{3}\dot{\rho},$$

and from the relation for conservation of energy-momentum of Eq. (10.14)

$$\dot{\rho} = -3\frac{\dot{a}}{a}\left(\rho + \frac{p}{c^2}\right) = -3H\left(\rho + \frac{p}{c^2}\right),$$

(recalling that $H = \dot{a}/a$), which yields from the above

$$2H\dot{H} = \frac{8\pi G}{3}\left(-3H\left(\rho + \frac{p}{c^2}\right)\right).$$

Dividing both sides by $2H$ ($\neq 0$)

$$\dot{H} = -4\pi G\left(\rho + \frac{p}{c^2}\right).$$

We need the expression for $\ddot{a}/a$ and we introduce this via the result of the previous Exercise†† in Chap. 10:

$$\dot{H} = \frac{\ddot{a}}{a} - H^2.$$

Then we can substitute this into the preceding equation so that

$$\left(\frac{\ddot{a}}{a} - H^2\right) = -4\pi G\left(\rho + \frac{p}{c^2}\right),$$

and consequently,

$$\frac{\ddot{a}}{a} = -4\pi G\left(\rho + \frac{p}{c^2}\right) + H^2 = \frac{-4\pi G}{3}\left(3\rho + \frac{3p}{c^2}\right) + H^2.$$

At this stage we can substitute in the value of H^2 from the first Friedmann equation,

$$\frac{\ddot{a}}{a} = \frac{-4\pi G}{3}\left(3\rho + \frac{3p}{c^2}\right) + \frac{8\pi G}{3}\rho + \frac{\Lambda c^2}{3},$$

and combining the ρ terms,

$$\frac{\ddot{a}}{a} = \frac{-4\pi G}{3}\left([3\rho - 2\rho] + \frac{3p}{c^2}\right) + \frac{\Lambda c^2}{3} = \frac{-4\pi G}{3}\left(\rho + \frac{3p}{c^2}\right) + \frac{\Lambda c^2}{3},$$

which is Eq. (10.13) as desired.

The calculation is more extensive if $k \neq 0$ and we leave that to the reader.

Terminology

All terms are defined in the context of Astronomy and may have slightly different interpretations in other scientific fields.

Absolute Magnitude	See Magnitude.
Apparent Magnitude	See Magnitude.
Apsidal Precession	The gradual rotation in the orientation of the orbit of a celestial body.
Apsis/Apsides (pl.)	The points in the orbital plane of a celestial body where it is closest to or furthest from the focal point of its elliptical orbit. The closest point is called the *periapsis*, and the furthest point is called the *apoapsis*.
Arcminute	One-sixtieth of a degree.
Arcsecond	One-sixtieth of an arcminute.
Argument of Periapsis (ω)	The angle measured in the reference plane from the ascending node to the point of closest approach (periapsis) of an orbiting body to the primary body. Called the *Argument of Perihelion* for Solar System bodies orbiting the Sun.
Ascending Node (Ω)	The point where an orbiting satellite passes through a reference plane from south to north.
Astronomical Unit	The mean distance of the Earth to the Sun equal to 149,597,870,700 km.
Barycenter	The common center of mass of two or more orbiting bodies and the point around which each of the bodies orbit.

Baryonic Matter	The type of matter composed of baryons (mainly protons and neutrons,) that make up the nuclei of atoms. Although electrons are not baryons, in cosmological contexts electrons are generally included so that baryonic matter is the same as ordinary matter. It comprises about 5% of the total mass-energy density of the Universe.
Big Bang	A scientific explanation for the origin of the Universe which postulates that the Universe began approximately 13.8 billion years ago from an extremely hot, dense state that rapidly expanded and cooled, leading to the formation of atoms, stars, and galaxies and other celestial structures.
Black Hole	A region of spacetime where the gravity is so intense that nothing, not even light can escape from it.
Blackbody	An idealized object that absorbs all incident electromagnetic radiation (with no reflection) irrespective of frequency/wavelength and emits radiation in a continuous spectrum with the intensity of radiation at each wavelength determined solely by its temperature, and not on its composition or surface properties.
Blackbody Radiation	The radiation that is emitted by a black body at all wavelengths having varying intensities whose peak intensity is determined by its temperature.
Blueshift	The decrease in the wavelength of electromagnetic radiation commonly observed in light from celestial objects moving toward the observer.
Brightness	A general term indicating the intensity of light that is perceived by an observer. In reference to stars it is defined in terms of the star's apparent brightness (as perceived from Earth) and intrinsic brightness (the amount of light energy emitted by the star measured as *luminosity*).
Celestial Equator	The Earth's equator projected out onto the celestial sphere forming a great circle.

Celestial Sphere	An imaginary sphere having the Earth at its center on which all other celestial bodies give the appearance of being projected at a similar distance.
Central Velocity Dispersion (σ)	Refers to the spread of velocities from the average of stars or gas within a celestial object such as a galaxy and is computed via the formula for standard deviation. It is often used to infer the mass of an unseen black hole at the core of a galaxy.
Centripetal Force	The force required to keep an object moving in a circular path and is directed inwards towards the center of rotation.
Cepheid	A type of variable star that exhibits periodic changes in its luminosity and are used as 'standard candles' in measuring cosmic distances.
Chandrasekhar Limit	Formulated by the theoretical physicist Subrahmanyan Chandrasekhar in 1930, it represents the maximum mass that a white dwarf star can have before collapsing due to its own gravity. It is equal to approximately 1.4 times the mass of the Sun. Exceeding the limit can lead to a neutron star, black hole, or Type 1a supernova.
Comoving Distance	The term refers to the distances between celestial objects that remains unchanged with time due to the expansion of the Universe (Hubble Flow).
Critical Density	The threshold density of the Universe that determines the overall geometry and fate of the Universe.
Dark Energy	A form of energy permeating all of space whose repulsive effect is driving the accelerated expansion of the Universe. Dark energy comprises about 68% of the mass-energy density of the Universe.
Dark Matter	An unknown form of matter than is inferred from the behavior of galaxy clusters and the rotation curves of stars and gas in spherical galaxies. Dark matter comprises about 27% of the mass-energy density of the Universe.
de Sitter Universe	A universe in which there is no ordinary mass and is completely dominated by dark energy embodied in the cosmological constant.

Distance Modulus	The quantity $\mu = m - M$ where m and M are the apparent and absolute magnitudes of a celestial object. It is directly related to the distance D of the object via the formula: $m - M = 5\log_{10}D - 5$.
Doppler Shift	A change in the frequency/wavelength of a wave due to the relative motion of the observer and light source. This includes both blueshift and redshift from objects moving either towards or away from the observer respectively. It is observed in all types of electromagnetic radiation.
Drake Equation	An equation formulated by astronomer Frank Drake that attempts to estimate how many civilizations exist in the Milky Way that are advanced enough to communicate with.
Eccentricity (e)	A parameter that defines the shape of an elliptical orbit. An eccentricity of zero defines a circle whereas a high eccentricity indicates a very elongated ellipse.
Ecliptic	The intersection of the plane of the Earth's orbit with the celestial sphere.
Effective Temperature	The temperature of a blackbody that corresponds to the same energy or luminosity given off by the star according to the Stefan-Boltzmann Law.
Einstein Field Equations	A set of ten interrelated equations of Einstein's General Theory of Relativity that describe how matter and energy in the Universe influence the curvature of spacetime, which is perceived as gravity.
Einstein Ring	A phenomenon whereby the light from a distant galaxy or star is distorted into the shape of a ring by the gravitational field of a massive body between the observer and distant light source. An exact alignment between the observer, intervening lensing object and the light source is needed to achieve a complete ring. Predicted by Einstein's General Theory of Relativity.
Ellipticity (E)	A measure of the deviation of an elliptical shape from that of a circle by comparing the semi-minor axis b to that of the semi-major axis a in the form $E = 1 - \frac{b}{a}$.

Entropy	A concept from thermodynamics that is a measure of how disorganized a system is. Higher entropy means higher disorder and natural systems tend to move towards a state of higher entropy. This is spelled out in The Second Law of Thermodynamics which states that the total entropy of a closed system never decreases over time.
Equation of State Parameter (w)	The quantity $w = p/\rho$, where p is the pressure of the substance and ρ is the energy density. In Cosmology, different values of this parameter lead to different evolutionary outcomes involving the fate of the Universe.
Equinox	One of the two points on the celestial sphere where the ecliptic crosses the celestial equator. The *vernal equinox* takes place on ~March 21 when the Sun crosses from South to North and the *autumnal equinox* is on ~23 September when the Sun crosses from North to South.
Escape Velocity	The velocity necessary to overcome the gravitational pull of a celestial body.
Event Horizon	Refers to the boundary in spacetime surrounding a black hole beyond which no light or other signals can escape due to the extreme gravitational pull. No matter or radiation crossing this boundary from the outside can escape the gravitational force of the black hole.
First point of Aries (♈)	The position of the Sun in the sky at the time of the vernal equinox. It is used as the zero point for measuring right ascension and celestial longitude.
FLRW (Friedmann-Lemaître-Robertson-Waker) Metric	A solution to the Einstein Field Equations widely used in Cosmology that describes a homogeneous, isotropic, expanding or contracting universe. It is used to model the large-scale structure and evolution of the Universe since the Big Bang.
Flux Density	See Irradiance.
Friedmann Equations	A set of equations deduced by Alexander Friedmann in 1922 from Einstein's Field Equations that model a homogeneous, isotropic universe.

General Relativity	The theory of gravitation developed by Albert Einstein and published in 1915, generalizing Special Relativity and Newton's Law of Universal Gravitation. It provided a unified description of gravity as a geometric consequence of the curvature of space and time (spacetime).
Global Positioning System (GPS)	A satellite-based navigation system consisting of 31 geosynchronous satellites, orbiting so that at least four are visible from any point on Earth and allow for the precise location of objects on Earth or in the atmosphere.
Gravitational Lensing	The phenomenon where the gravity of a massive celestial object bends and magnifies the light from a distant light source. The effect was predicted by Einstein's General Theory of Relativity.
Gravitational Redshift	A phenomenon in which light or other electromagnetic radiation emanating from a massive object is shifted towards the longer wavelength, or redshifted, as it climbs out of the gravitational well of the object. One of the many consequences of the General Theory of Relativity.
Hertzsprung-Russell Diagram	A plot of the absolute magnitude (luminosity) of stars versus their spectral class (effective temperature). It was developed by Einar Hertzsprung and H.N. Russell early in the twentieth century.
Holographic Principle	The notion that all the information about a volume of space can be thought of as encoded on its boundary. The idea was explored by Gerard 't Hooft and Leonard Susskind and has been very influential in Theoretical Physics.
Homogeneous	Viewed on the largest scales (millions of light-years), the average distribution of matter is the same throughout the Universe.
Hubble Constant (H_0)	The parameter in Cosmology that describes the rate at which the Universe is expanding, quantifying the distance to a galaxy with its velocity of recession. The constant has a value between 67 to 74 (km/s)/Mpc.

Hubble Law	The mathematical expression showing that the velocity of recession of a galaxy is proportional to its distance via the formula: $v_r = H_0 D$, where H_0 is the Hubble Constant and D is the distance to the galaxy.
Inclination (i)	The tilt of the plane of an orbiting body relative to a reference plane. For Solar System bodies, the reference plane is the plane of the Earth's orbit around the Sun, known as the *ecliptic plane* or *plane of the ecliptic*.
Irradiance (Flux/Radiant Flux density)	The power of electromagnetic radiation incident on a surface per unit area typically measured in W/m².
Isotropic	Viewed on large scales (millions of light-years), the structure of the Universe appears the same in every direction.
Jeans Criterion (Instability)	The conditions under which an interstellar cloud of gas will begin to collapse under its own gravitational force possibly leading to the formation of a star.
Julian Year	The length a year according to the Julian calendar and equal to 365.25 days. It is often used in astronomical calculations.
Lagrange Points	The five points in space where the gravitation effects from two massive orbiting bodies are such that a small body at these points tends to remain in a stable or quasi-stable relation with the larger bodies.
Law of Universal Gravitation	The statement that every particle in the Universe attracts every other particle with a force that is proportional to the product of their masses and inversely proportional to the distance between their centers. It was enunciated by Isaac Newton in his *Principia* of 1687.
Light Cone	A cone-shape diagram from Special Relativity that represents all possible paths that light can take from a specific event through spacetime. The cone splits into two sections: the future light cone and the past light cone.
Light-Year	The distance that light travels in one Julian year.

Longitude of the Ascending Node (Ω)	The angle measured from the reference direction (vernal equinox) to the point (ascending node) where an orbiting body has passed from south to north through the reference plane.
Lookback Time	The concept of observing distant celestial objects as they were in the past due to the finite speed of light and the time it takes to reach us. It is the time it has taken for the light to reach us from when it was emitted. Thus, light from a galaxy one-billion light-years away shows us what the galaxy looked like one billion years ago.
Lorentz Dilation Factor	In Special Relativity, when an object moves at a velocity approaching the speed of light relative to a stationary observer, then time for the moving object slows down in accordance with the Lorentz factor: $\gamma = 1/\sqrt{1 - \frac{v^2}{c^2}}$, where v is the velocity of the moving object and c is the velocity of light.
Luminosity	The total amount of electromagnetic radiation emitted across all wavelengths by an object per unit of time. It is an intrinsic property of the object irrespective of distance. The units are in watts (W).
Magnitude	The logarithmic scale used to describe the brightness of an object as seen from an observer on Earth. The *apparent magnitude* is a measure of the brightness of an object as seen from Earth, whereas the *absolute magnitude* is a measure of the intrinsic brightness of the object as if it were located at a distance of 10 parsecs.
Mean Anomaly	The angle that arises from converting the fraction of an orbiting body's elapsed time since that of periapsis divided by the orbital period.
Metric Tensor	A matrix with variable entries that allows for the calculation of distances in spacetime by taking into account the curvature at each point.

Terminology

Microlensing	An astronomical phenomenon occurring when the gravitational field of a massive celestial object bends the light from a more distant background object. This results in an increase in a temporary increase in brightness as the lens passes in front of the source as viewed from Earth.
Parallax	The apparent angular displacement of a celestial object against the background stars when observed from two different points in the Earth's orbit.
Parsec	The distance to an object when the 1 AU radius of the Earth's orbit subtends an angle of 1 arcsec. It is equal to 3.26 light-years.
Peculiar Motion	The movement of a celestial body that has the Hubble flow factored out. Is a consequence of the gravitational effects of other nearby bodies.
Period-Luminosity Relation	The relationship between the pulsation period of a Cepheid and the luminosity of the star. A longer period is associated with greater luminosity in a quantifiable manner.
Planck's Law	A mathematical description of the intensity of radiation emitted by a blackbody at different wavelengths λ and at different temperatures T for the spectral radiance.
Power Law	A relation of the form $y = cx^{\alpha}$ for some constant c and power α ($x > 0$), that is found in the behavior of many natural phenomena. Power laws are scale invariant in that a change of scale of the variable x results in the proportional scaling of y so that the shape of the relation remains the same.
Quintessence	A proposed formulation of dark energy characterized by a dynamic scaler field that varies over time (and potentially space), leading to a time-dependent energy density and evolving equation of state.
Radian	The angle that is subtended by the length of the radius placed on the circumference of a circle.

Radiance	Measures the power emitted or received per unit solid angle per unit area. It describes how much light is received from a specific direction. The units are $W \cdot sr^{-1} \cdot m^{-2}$.
Radiant Flux	The total amount of energy received by an object per unit of area per unit of time. It is the total power received from a celestial body per unit area measured by an observer. It obeys an inverse square law and its units are watts per square meter (W/m^2).
Raleigh-Jeans Law	An approximation of Planck's Law for large values of the wavelength λ. Its poor approximation for small wavelength values leads to the Ultraviolet Catastrophe.
Redshift	The increase in the wavelength of electromagnetic radiation due to the movement of the source away from the observer or due to gravity.
Regression Line	The straight line found by the method of least squares that 'best' represents a set of data points.
Roche Limit	The minimum distance that a satellite can remain in orbit around a primary body without being torn apart by tidal forces.
Scale Factor ($a(t)$)	A quantity that relates the changes in the distance over time between two points in the Universe. If $D(t_0)$ is a reference distance (say, at the present time t_0), then the distance at time t satisfies the relation: $D(t) = a(t)D(t_0)$.
Schwarzschild Radius	The radius of the event horizon for a nonrotating, uncharged (Schwarzschild) black hole given by $R_S = \frac{2GM_{BH}}{c^2}$.
Semi-Major Axis	The distance of one-half the longest diameter of an ellipse.
Semi-Minor Axis	The distance of one-half the shortest diameter of an ellipse.
Sérsic Profile	A function that describes how the intensity of light emitted from a celestial object, such as a galaxy, varies with the distance from its center.
Sidereal Year	Time that the Earth takes to complete one orbit of the Sun with respect to the fixed background stars. It is equal to 365.25636 days.

Terminology

Singularity — In the context of General Relativity, it is a point at the center of a black hole characterized by infinite curvature and density. It indicates that our understanding of spacetime and gravity is incomplete.

Spacetime — The weaving together of the three dimensions of space with one dimension of time into a four-dimensional continuum that is the framework for describing how mass and energy behave in the Universe.

Special Relativity — Developed by Albert Einstein in 1905, it was a new framework for dealing with space and time without considering the effects of gravity. It is especially important when discussing the behavior of objects moving at close to the speed of light and also led to the famous equation: $E = mc^2$, published in the same year.

Spectral Irradiance — The power of electromagnetic radiation that is received per unit area, per unit wavelength from all directions. It is in units of $W \cdot m^{-2} \cdot nm^{-1}$. It is used to study the composition and temperature, distance and size of stars.

Spectral Radiance — The amount of power emitted or reflected per unit steradian, per unit area, per unit wavelength ($W \cdot sr^{-1} \cdot m^{-2} \cdot nm^{-1}$). It is used for analyzing the distribution of energy emitted or reflected by an object across different wavelengths which provides information about the object's temperature and composition.

Standard Candle — A celestial object that has a known luminosity and so can be used in order to measure distances by comparing the luminosity (absolute magnitude) with its apparent magnitude via the inverse square law.

Standard Deviation — A statistical measure that quantifies the amount of variability or dispersion in a set of data values indicating how much individual data points deviate from the mean.

Stefan-Boltzmann Law — A formula establishing that the power radiated (in W/m^2) by a blackbody is proportional to the fourth power of its temperature (in K). The constant of proportionality $\sigma = 5.670374 \times 10^{-8} W/m^2 \cdot K^4$ is the *Stefan-Boltzmann constant*.

Steradian	The solid angle formed by the origin of a sphere of radius r that subtends a circular area on the sphere equal to the square of the radius.
Tidal Force	The differential gravitational pull exerted by one body on the different parts of another due to the differences in distance from the source of the gravity.
Time Dilation	A phenomenon predicted by the Theory of Relativity whereby time appears to elapse at different rates in different reference frames, particularly those with high velocity and/or high gravitational fields.
Trilateration	A method for determining a specific location in space knowing the distances from it to three or more reference points.
True Anomaly (ν)	The angular displacement between the direction of periapsis and the position of a satellite body in its orbit at a given time.
Tully-Fisher Relation	An empirical relation that links the luminosity of a spiral galaxy to its rotational velocity, expressed mathematically as $L \propto V_{max}^4$ where L is the luminosity and V_{max} is the maximum rotational velocity in the galaxy's outer regions.
Type 1a Supernova	A particular type of stellar explosion that occurs in a binary system where a white dwarf acquires sufficient mass from a companion to exceed the Chandrasekhar limit which triggers a runaway nuclear reaction. The light curve has a distinctive shape so that it can be used as a standard candle for measuring distances.
Ultraviolet Catastrophe	An erroneous consequence resulting from the Raleigh-Jeans Law for spectral radiance due to the law's inability to model small wavelengths accurately.
Universe	Refers to the totality of all space, time, matter, and energy that exist, including all the physical laws and constants that govern them. The observable is about 46.5 billion light-years in radius from any point.
Vernal Equinox	See equinox.

Virial Theorem	A basic principle stating that for a system in a steady state or equilibrium, the total kinetic energy K of the system is related to the total potential energy of the system U by the formula: $2K + U = 0$.
Watts	A measure of power in terms of joules per second (J/s).
Wavelength	The distance from one peak (or trough) to another in a periodic wave.
Wien's Approximation Law	A simplified version of Planck's Law which is a close approximation to it especially for shorter wavelengths.
Wien's Displacement Law	The inverse relation of the maximum wavelength of blackbody radiation for the body at a given temperature. It is given by: $\lambda_{max} = b/T$, where $b = 0.0028977$ (m $\cdot$ K) is Wien's constant and T is the temperature measured in Kelvin.

Further Reading

E.A. Beet, *Mathematical Astronomy for Amateurs*, Norton, 2010.
B.W. Carroll and D.A. Ostlie, *An Introduction to Modern Astrophysics*, 2nd Ed., Cambridge University Press, 2017.
D. Fleisch and J. Kregenow, *A Student's Guide to the Mathematics of Astronomy*, Cambridge U. Press, 2013.
N. Grossman, *The Sheer Joy of Celestial Mechanics*, Birkhäuser, 1996.
Inglis, M., *Astrophysics is Easy! An Introduction for the Amateur Astronomer*, Springer 2015.
Maoz, D. *Astrophysics in a Nutshell*, 2nd Ed., Princeton, 2016
Schiff, J.L., *The Most Interesting Galaxies in the Universe*, IOP Concise Physics, 2018.
Schiff, J.L., *The Mathematical Universe – From Pythagoras to Planck*, Springer-Praxis, 2020.

Index

A
Absolute magnitudes, 3, 43, 44, 107–117
Age of the Earth, 23–28
Angles, 4–6, 8, 14, 24, 36, 75–77, 80, 85, 102–105, 119, 134, 154, 155, 177, 180, 187, 196, 212, 221
Apparent magnitude, 3, 13, 14, 107, 108, 112, 114–117
Apsidal precession, 145–148, 220
Arcminute, 4
Arcseconds, 4–6, 85, 103, 120, 145–147, 154
Argument of periapsis, 77
Ascending node, 76, 77
Astronomical unit (AU), 3, 9, 83, 85, 104

B
Barycenter, 81
Baryonic matter, 187, 198
Big Bang, 11, 35, 96, 130, 173, 180, 182, 187, 188, 194, 195, 199
Blackbody, 35–37, 39, 40, 172, 212
Blackbody radiation, 34–41
Blueshift, 121
Boltzmann Constant, 3, 37, 40, 61, 174
Brightness, 12–14, 42, 44, 47, 49, 51, 105–110, 113, 156

C
Celestial equator, 75
Celestial sphere, 10
Central velocity dispersion, 166–168
Centripetal force, 78–81, 90
Cepheid variables, 43, 106, 107, 109–112, 120, 127
Chandrasekhar limit, 113, 151–153
Comoving, 195, 196
Critical density, 185–187

D
Dark energy, 176, 177, 179, 180, 182–184, 187, 188, 194, 197–199, 224
Dark matter, 60, 91–99, 158, 179, 180, 187, 188, 198
Density, 10–11, 42–43, 52, 60, 63, 67, 68, 92, 95, 96, 114, 162, 168–169, 176–180, 182–188, 194, 197, 198, 214, 215
de Sitter Universe, 182, 194
Distance formula, 101–102
Distance modulus, 110–114, 117, 128, 130, 164
Doppler shift, 115, 121–123, 126
Drake equation, 201–204

E
Eccentricity, 49, 70–74, 146, 170
Ecliptic, 74, 75
Effective temperature, 21, 40
Einstein field equations (EFE), 175–177, 179, 192, 195
Einstein ring, 156, 160, 221–223
Electromagnetic radiation, 1, 31–35, 121, 134, 136, 148, 161
Ellipticity, 73–74

Energy, 2, 20, 34, 37, 40, 41, 57–61, 64, 98, 122, 126, 148, 175–180, 182–188, 190, 191, 194, 197, 198, 214–215, 227
Entropy, 173–174, 198
Equation of state parameter, 198–199
Equinox, 75, 76
Escape velocity, 29, 64, 193
Event horizon, 88, 161–164, 170, 174, 193, 194

F
Flat space, 133
FLRW metric, 195
Flux, 41–43
Frequency, 32–35, 37–40, 121, 123, 148, 149, 212
Friedmann equations, 177–180, 182–185, 198, 224, 227–228

G
General relativity, 124, 125, 133, 141–142, 145–148, 150, 151, 153, 155, 156, 191
Geometry of space, 177–178
Geostationary satellites, 30, 89–90
Global Positioning System (GPS), 144–145, 218, 219
Gravity, 29, 41, 46, 54–60, 62, 64, 66, 79, 81, 90, 95, 96, 114, 121, 133, 136, 141, 142, 144, 145, 148, 153, 154, 161, 162, 170, 172, 173, 175, 190, 194, 195, 219, 223, 224
Gravitational potential, 57–59, 61, 64, 214, 215
Gravitational lensing, 96, 97, 99, 158, 223
Gravitational redshift, 148–160, 175

H
Hawking radiation/temperature, 170–172
Hertzsprung-Russell Diagram, 43–45, 114
Holographic principle, 174
Homogeneous, 175, 195
Hubble constant, 127, 128, 130, 185, 186, 189
Hubble Law, 129, 132

I
Inclination, 14, 15, 75, 76
Inverse square relation, 41–42
Irradiance, 42–43
Isotropic, 175, 195

J
Jeans Criterion, 60
Julian Day/Year, 3, 4

K
Kepler's Laws, 14, 69–99, 166

L
Lagrange Points, 90–91
Law of Universal Gravitation, 29, 54, 78
Light cone, 190, 191
Light-years, 3, 9–10, 50, 85, 97, 98, 101, 104, 105, 111–113, 120, 126, 132, 150, 151, 158, 164–166, 169, 186, 223
Logarithms, 12, 19, 25, 44, 48
Longitude of the ascending node, 76
Lookback time, 129, 130
Lorentz dilation factor, 140–141
Luminosity, 3, 20, 40, 41, 43–47, 109, 114–117, 164, 172

M
M87, 114, 164–168
Magnitude, 12, 13, 44, 47, 53, 56, 78, 106–108, 110, 111, 158, 224, 226
Main-sequence lifetimes, 21–22
Mass, 2, 27, 44, 53, 78, 106, 141, 161, 175
Mass-luminosity relation, 20, 21
Mass of Earth, 3, 29–30
Mean anomaly, 77
Method of least squares, 15–18, 205–206
Metric tensor, 191–193
Micro-arcsecond, 5
Microlensing, 52, 156–158
Milli-arcsecond, 5

N
Newton's Laws of motion, 53–54

O
Orbital elements, 74–77

P
Parallax, 10, 102–105, 107, 109
Parsecs, 3, 5, 9–10, 43, 104, 105, 107–110, 115, 117, 127, 128

Peculiar velocity, 196
Periapsis, 71, 72, 77, 83, 86
Period-luminosity relation, 107, 109–110
Planck's Law, 36–40, 208–211
Power Laws, 19–22, 115

Q
Quintessence, 199

R
Radiance, 36, 47
Radians, 4–8, 77, 80, 103, 104, 118–120, 146, 148, 155, 170
Radiant flux, 41–43
Raleigh-Jeans Law, 38, 39
Redshifts, 114, 118, 121–125, 127–130, 132, 148–158, 165, 166, 175, 176
Reduced Planck Constant, 3, 34, 152
Regression line, 15–17, 27, 28, 109, 110
Roche limit, 64–68
Rotational velocities, 14, 15, 90, 93, 102, 115–116

S
Sagittarius A*, 84–89, 166
Scale factors, 130, 132, 177–179, 182, 183, 194–196, 198, 226, 227
Schwarzschild precession, 147, 148, 169, 170
Schwarzschild radius, 96, 162–163, 167–170, 176, 193, 194
Scientific notation, 1–2
Semi-major axis, 69, 71, 72, 77, 80, 82–89, 146, 147, 170
Semi-minor axis, 69
Sérsic profile, 47–50
Sidereal Day/Year, 3, 4
SI units, 2
Size of the Earth, 24
Solid angles, 6–8, 212
Spacetime metrics, 189–191

Special relativity, 124, 125, 133–135, 190, 191
Spectral irradiance, 42
Spectral radiance, 36, 37, 39
Standard deviations, 17–18, 166
Stefan-Boltzmann constant, 3, 40
Stefan-Boltzmann Law, 40–41, 43, 212–214
Stellar magnitudes, 12–14
Steradian, 6–8, 37, 212

T
Temperatures, 2, 3, 11, 21, 35–40, 43, 45, 60, 61, 63, 114, 170–172, 174, 187, 197
Tidal forces, 64–68
Time dilation, 135–140, 142–144, 155, 175, 193, 194, 216–219
Tip of the red giant branch (TRGB), 114, 129, 164, 165, 189
Topology of the Universe, 196–197
Transit method, 50–52
True anomaly, 77
Tully-Fisher relation, 43, 115–118
Type 1a supernovae, 113–114, 129, 189

U
Ultraviolet catastrophe, 39–40
Universe, 9, 31, 54, 77, 113, 121, 141, 161, 175, 201

V
Vernal equinox, 75
Virial Theorem, 59–64, 97–99

W
Watts, 2, 20, 40
Wavelengths, 20, 31–40, 42, 47, 121–123, 126, 132, 148, 149, 212
Wien's Approximation Law, 37, 211
Wien's Displacement Law, 35–36, 208–211

MIX
Papier aus verantwortungsvollen Quellen
Paper from responsible sources
FSC® C105338

If you have any concerns about our products,
you can contact us on
ProductSafety@springernature.com

In case Publisher is established outside the EU,
the EU authorized representative is:
Springer Nature Customer Service Center GmbH
Europaplatz 3, 69115 Heidelberg, Germany

Printed by Libri Plureos GmbH
in Hamburg, Germany